Correlated

CALGARY PUBLIC LIBRARY

JAN   2015

Correlated

→ **SURPRISING CONNECTIONS** ←
BETWEEN SEEMINGLY UNRELATED THINGS

Shaun Gallagher

A PERIGEE BOOK

A PERIGEE BOOK
Published by the Penguin Group
Penguin Group (USA) LLC
375 Hudson Street, New York, New York 10014

USA • Canada • UK • Ireland • Australia
New Zealand • India • South Africa • China

penguin.com

A Penguin Random House Company

CORRELATED

Copyright © 2014 by Shaun Gallagher
Penguin supports copyright. Copyright fuels creativity, encourages diverse voices, promotes free speech, and creates a vibrant culture. Thank you for buying an authorized edition of this book and for complying with copyright laws by not reproducing, scanning, or distributing any part of it in any form without permission. You are supporting writers and allowing Penguin to continue to publish books for every reader.

PERIGEE is a registered trademark of Penguin Group (USA) LLC.
The "P" design is a trademark belonging to Penguin Group (USA) LLC.

ISBN: 978-0-399-16247-3

An application to catalog this book has been submitted to the Library of Congress.

First edition: July 2014

PRINTED IN THE UNITED STATES OF AMERICA

10 9 8 7 6 5 4 3 2 1

While the author has made every effort to provide accurate telephone numbers, Internet addresses, and other contact information at the time of publication, neither the publisher nor the author assumes any responsibility for errors, or for changes that occur after publication. Further, the publisher does not have any control over and does not assume any responsibility for author or third-party websites or their content.

Most Perigee books are available at special quantity discounts for bulk purchases for sales promotions, premiums, fund-raising, or educational use. Special books, or book excerpts, can also be created to fit specific needs. For details, write: Special.Markets@us.penguingroup.com.

To all those who have joined in the fun at Correlated.org.

INTRODUCTION

If you've ever wondered whether iPod owners are more likely to stir their drinks counterclockwise, whether nonfiction lovers are more likely to have a positive opinion about France, or whether tea drinkers are more likely to prefer mechanical pencils, then this is the book for you!

The data that underpin the statistics you'll find in this book come from Correlated.org, a website devoted to uncovering surprising correlations between seemingly unrelated things.

Each day, a new poll question is posted on the site, and at the end of the day, the poll responses are tallied up and compared with the results of every previous poll to find the two responses with the strongest correlation.

For this book, I've taken Correlated.org's large data set—1,089,173 poll responses from 36,305 respondents—and generated all-new statistics in which the results of each poll were compared not only with previous polls but with the entirety of the available data. For the 182 topics chosen for inclusion in the book, the median sample size was 2,290 and the mean sample size was 2,222.

Although the statistics in this book are all based on

real data, the methodology by which the correlations are generated is not intended to stand up to professional scrutiny. (In fact, it would probably make a professional statistician weep.) Rather, the correlations are intended to be amusing and thought provoking, and to illustrate some of the absurdities that result when you ignore tricky concepts such as confounding variables or the multiple-testing problem.

I hope you enjoy the surprising statistics presented here, and I hope you'll join the thousands of others who help generate new correlations by contributing to Correlated.org.

TOPICS

'80s Music

40% of people have boycotted a company.

But among those who consider music from the '80s to be oldies, **52%** have boycotted a company.

53% of people are Foo Fighters fans.

But among those who consider music from the '80s to be oldies, only **41%** are Foo Fighters fans.

70% of people are adept at using chopsticks.

But among those who consider music from the '80s to be oldies, only **58%** are adept at using chopsticks.

DOUBLE PLAY

You're extremely likely to consider music from the '80s to be oldies if you both:

- always wash your hands after using a public restroom
- prefer Jake Gyllenhaal's movies to Maggie Gyllenhaal's

NO CORRELATION

People who consider music from the '80s to be oldies are almost exactly as likely as the average person to:

- have been involved in student government
- dislike honey-mustard sauce
- have a nine-to-five type of job

AAA (The Auto Club)

38% of people have had a poison ivy rash.

But among those who belong to AAA, **54%** have had a poison ivy rash.

77% of people say that if a person dislikes them, it's generally because they don't know "the real me."

But among those who belong to AAA, **89%** say that if a person dislikes them, it's generally because they don't know "the real me."

29% of people have visited a chiropractor.

But among those who belong to AAA, **40%** have visited a chiropractor.

DOUBLE PLAY
You're extremely likely to belong to AAA if you both:
- prefer a hot breakfast over a cold breakfast
- consider yourself more generous than the average person

NO CORRELATION
People who belong to AAA are almost exactly as likely as the average person to:
- prefer Google Chrome as their browser
- say they'd agree to work a midnight-to-8-a.m. shift for the next 10 years if the pay were $1 million a year
- have never had asthma

Absent-Mindedness

25% of people say they are better at punctuality than punctuation.

But among those who wouldn't describe themselves as absent-minded, **38%** say they are better at punctuality than punctuation.

32% of people say they had a negative first impression of Pope Francis.

But among those who wouldn't describe themselves as absent-minded, **44%** had a negative first impression of Pope Francis.

42% of people regularly eat yogurt.

people who think New Yorkers are cooler than Los Angelenos

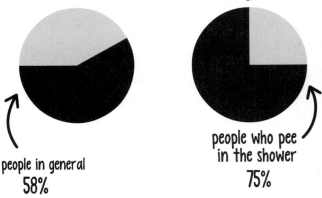

people in general
58%

people who pee
in the shower
75%

But among those who wouldn't describe themselves as absent-minded, **52%** regularly eat yogurt.

DOUBLE PLAY

You're extremely likely to say you aren't absent-minded if you both:

- aren't a coward
- think clowns are fun

NO CORRELATION

People who wouldn't describe themselves as absent-minded are almost exactly as likely as the average person to:

- say the city where they live is more important than the job they do
- keep it a secret if they found out they had only six months to live
- have participated in an eating contest

Adam and Eve

38% of people ascend the stairs two at a time.

But among those who don't think Adam and Eve had belly buttons, **50%** ascend the stairs two at a time.

36% of people dislike sideburns.

But among those who don't think Adam and Eve had belly buttons, **48%** dislike sideburns.

49% of people prefer *Steven* over *Stephen*.

But among those who don't think Adam and Eve had belly buttons, **60%** prefer *Steven* over *Stephen*.

DOUBLE PLAY

You're extremely likely to think that Adam and Eve didn't have belly buttons if you both:

- find Australians sexy
- are not a fan of Coldplay

NO CORRELATION

People who don't think Adam and Eve had belly buttons are almost exactly as likely as the average person to:

- refuse to tip a server when they receive poor service
- prefer to be the X in tic-tac-toe
- own flat-screen TVs

Air Freshener

66% of people like cotton candy.

But among those who use air freshener in their home, **79%** like cotton candy.

45% of people say they are closer to lower class than upper class.

But among those who use air freshener in their home, **58%** say they are closer to lower class than upper class.

42% of people have bouts of indigestion more often than bouts of indignation.

But among those who use air freshener in their home, **52%** have bouts of indigestion more often than bouts of indignation.

DOUBLE PLAY
You're extremely likely to use air freshener in your home if you both:
- prefer debit cards over credit cards
- regularly wear cologne/perfume

NO CORRELATION
People who use air freshener in their home are almost exactly as likely as the average person to:
- fear public speaking
- apply antiperspirant at night
- put a cap on uppercase Js

Aisle Seat vs. Window Seat

34% of people tend to micromanage when they're in a leadership role.

But among those who prefer an aisle seat on a plane, only **20%** tend to micromanage when they're in a leadership role.

41% of people aren't bothered by parents who kiss their adult children on the lips.

But among those who prefer an aisle seat on a plane, **54%** aren't bothered by parents who kiss their adult children on the lips.

STATISTICS 101: MARGIN OF ERROR

Suppose you live in a bustling city with half a million residents, and you want to know what percentage of those residents prefer Beethoven to Mozart.

It would be impractical to ask all 500,000 of them, so instead, you decide to ask a representative sample.

As you might imagine, the larger your sample size, the more likely their responses will reflect the preferences of the entire population.

A poll's margin of error tells you the range of expected variation between your poll results and the value you would get if you polled the entire population. Margin of error decreases as your sample size grows.

Here's an example: Suppose you poll 1,000 residents and find that 48% prefer Beethoven and 52% prefer Mozart.

With about 95% confidence (see page 19), your margin of error is plus or minus 3 percentage points, meaning it's highly likely that if you were to repeat your survey with another sample group, the percentage of people who say they prefer Beethoven would be between 45% and 51%, and the percentage of people who say they prefer Mozart would be between 49% and 55%.

It's plausible, however, that if you were to poll the entire city, you might find the percentage of people who prefer Beethoven is actually at the top of the expected range, 51%, and the percentage who prefer Mozart might be at the bottom of the expected range, 49%.

Because of this possibility, when the range of values that fall within a poll's margin of error overlap, the results of the poll are often described as a statistical tie, meaning that we cannot be sufficiently confident that one option is preferred over the other.

39% of people believe the chicken came before the egg.

But among those who prefer an aisle seat on a plane, **52%** believe the chicken came before the egg.

DOUBLE PLAY

You're extremely likely to prefer an aisle seat on a plane if you both:

- say your primary alarm clock is not your phone
- are good at parallel parking

NO CORRELATION

People who prefer an aisle seat on a plane are almost exactly as likely as the average person to:

- prefer to dry their hands with paper towels in public restrooms
- say they can perform a handstand
- cut the fabric tags off of clothing

Aliens

60% of people have never been in a fistfight.

But among those who don't believe aliens exist, **72%** have never been in a fistfight.

49% of people swear a lot.

But among those who don't believe aliens exist, only **37%** swear a lot.

32% of people have had to make a life-or-death decision.

But among those who don't believe aliens exist, only **21%** have had to make a life-or-death decision.

DOUBLE PLAY
You're extremely likely to think aliens don't exist if you both:
- oppose gay marriage
- think prostitution should be illegal

NO CORRELATION
People who don't believe aliens exist are almost exactly as likely as the average person to:
- have never cried over the death of someone they've never met
- prefer little dogs
- not own any leather clothing, aside from footwear or belts

All-Nighters

45% of people use the phrase *a couple* to mean "a few," rather than exactly two of something.

But among those who have never pulled an all-nighter, **67%** use the phrase *a couple* to mean "a few," rather than exactly two of something.

64% of people think that fooling kids into thinking that Santa is real is a harmless fib.

But among those who have never pulled an all-nighter, **82%** think that fooling kids into thinking that Santa is real is a harmless fib.

20% of people prefer their application dock/dashboard to run along the side of their screen.

But among those who have never pulled an all-nighter, **37%** prefer their application dock/dashboard to run along the side of their screen.

DOUBLE PLAY
You're extremely likely to have never pulled an all-nighter if you both:
- tilt your head to the left when you go in for a kiss
- don't have a long commute

NO CORRELATION
People who have never pulled an all-nighter are almost exactly as likely as the average person to:
- be dog lovers
- regularly clip coupons
- be more likely to splurge on where they stay than where they eat while on vacation

Bananas

53% of people prefer lemonade to iced tea.

But among those who don't like bananas, **71%** prefer lemonade to iced tea.

74% of people like pumpkin pie.

But among those who don't like bananas, only **57%** like pumpkin pie.

24% of people say they have curly hair.

But among those who don't like bananas, **39%** have curly hair.

DOUBLE PLAY

You're extremely likely to dislike bananas if you both:

- think of soup as only a cold-weather food
- don't like guacamole

NO CORRELATION

People who don't like bananas are almost exactly as likely as the average person to:

- think a moat is a more important feature in a castle than secret passages
- prefer a neat-freak roommate over a messy one
- be good at remembering the words to songs

Best Friends

38% of people regularly use an RSS reader.

But among those who don't have a best friend, **51%** regularly use an RSS reader.

51% of people say their favorite kind of melon is not watermelon.

But among those who don't have a best friend, **63%** say their favorite kind of melon is not watermelon.

74% of people prefer Alex P. Keaton of *Family Ties* to Mike Seaver of *Growing Pains*.

But among those who don't have a best friend, **85%** prefer Alex P. Keaton of *Family Ties* to Mike Seaver of *Growing Pains*.

DOUBLE PLAY
You're extremely likely to not have a best friend if you both:
- aren't interested in sign language
- trust professional critics' movie recommendations more than those of your friends

NO CORRELATION
People who don't have a best friend are almost exactly as likely as the average person to:
- say their political opinions are substantially different from those of their parents
- rather see increased funding to the arts than to the sciences
- like white pizza

Big Government vs. Big Business

23% of people think climate change is not caused by humans.

But among those who dislike Big Government more than Big Business, **49%** think climate change is not caused by humans.

49% of people think a woman should take her husband's last name.

But among those who dislike Big Government more than Big Business, **68%** think a woman should take her husband's last name.

54% of people have a neutral or positive opinion about homeschooling.

But among those who dislike Big Government more than Big Business, **66%** have a neutral or positive opinion about homeschooling.

DOUBLE PLAY

You're extremely likely to dislike Big Government more than Big Business if you both:

- oppose embryonic stem cell research
- are confused by modern art

people who think we're too stingy with foreign aid

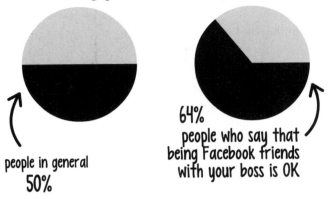

people in general
50%

64%
people who say that
being Facebook friends
with your boss is OK

NO CORRELATION

People who dislike Big Government more than Big Business are almost exactly as likely as the average person to:

- like marshmallows in their hot chocolate
- take pleasure in putting things in order
- say they'd rather ride a roller coaster than a water-slide

Black Friday

25% of people find QR codes useful.

But among those who join in the Black Friday shopping frenzy, **44%** find QR codes useful.

46% of people think fireworks displays are better when they're set to music.

But among those who join in the Black Friday shopping frenzy, **62%** think fireworks displays are better when they're set to music.

56% of people would be more likely to be police officers than firefighters.

But among those who join in the Black Friday shopping frenzy, **70%** would be more likely to be police officers than firefighters.

DOUBLE PLAY

You're extremely likely to join in the Black Friday shopping frenzy if you both:

- are more likely to sleep on your left side
- would rather host the Academy Awards than the Grammys

NO CORRELATION

People who join in the Black Friday shopping frenzy are almost exactly as likely as the average person to:

- have collected unemployment
- think sunrise is more beautiful than sunset
- be content with their present financial situation

Blind Dates

39% of people have business cards.

But among those who have been on blind dates, **56%** have business cards.

38% of people have been cheated on.

But among those who have been on blind dates, **54%** have been cheated on.

49% of people prefer *Steven* over *Stephen*.

But among those who have been on blind dates, **65%** prefer *Steven* over *Stephen*.

DOUBLE PLAY
You're extremely likely to have been on a blind date if you both:
- are a teacher
- have served as a best man or maid of honor

NO CORRELATION
People who have been on blind dates are almost exactly as likely as the average person to:
- prefer sprinkles on their ice cream cones
- wash their own cars
- prefer cushioned toilet seats

Blind vs. Deaf

21% of people write poetry.

But among those who would rather be blind than deaf, **36%** write poetry.

32% of people can play a stringed instrument.

But among those who would rather be blind than deaf, **47%** can play a stringed instrument.

75% of people aren't picky about foods touching each other on their dinner plate.

But among those who would rather be blind than deaf, **88%** aren't picky about foods touching each other on their dinner plate.

DOUBLE PLAY
You're extremely likely to prefer being blind to being deaf if you both:

- describe yourself as a good singer
- prefer *okay* over *OK*

NO CORRELATION
People who would rather be blind than deaf are almost exactly as likely as the average person to:

- dislike their neighbors
- prefer ibuprofen to acetaminophen for minor pain
- like Jerry Seinfeld

Blood

42% of people are afraid of heights.

But among those who are disturbed by the sight of blood, **57%** are afraid of heights.

64% of people get more fired up about politics than about religion.

But among those who are disturbed by the sight of blood, **77%** get more fired up about politics than religion.

35% of people have donated to a political campaign.

But among those who are disturbed by the sight of blood, **48%** have donated to a political campaign.

DOUBLE PLAY
You're extremely likely to be disturbed by the sight of blood if you both:
- prefer little dogs to big dogs
- say that your ability to play sports is lower than your interest in sports

NO CORRELATION
People who are disturbed by the sight of blood are almost exactly as likely as the average person to:
- like butter on their movie popcorn
- be up to date on their tetanus shots
- like spicy food

STATISTICS 101: CONFIDENCE LEVELS

When the results of a poll are released, pollsters ordinarily include the margin of error (see page 7), which depends on a concept called *statistical confidence*.

Basically, the more confident you want to be in your results, the greater margin of error you need to allow.

For instance, suppose a poll of 1,000 people shows that 45% admire Teddy Roosevelt more than Franklin D. Roosevelt. Using a confidence level of 95%, we can determine that the margin of error is plus or minus 3%.

The confidence level of 95% means that if you were to repeat your poll 20 times, you would expect the results to fall within the margin of error in 19 of 20 cases (95% of the time).

Although pollsters ordinarily use a confidence level of 95%, it's possible to use other values, such as a more stringent 99%. In the example here, if you want to be 99% confident that your margin of error is wide enough, you'd need a margin of error of plus or minus 4%.

Body Piercings

53% of people like punk rock.

But among those with body piercings, **82%** like punk rock.

35% of people get stressed about the holidays.

But among those with body piercings, **62%** get stressed about the holidays.

24% of people have deployed a fire extinguisher.

But among those with body piercings, **50%** have deployed a fire extinguisher.

DOUBLE PLAY
You're extremely likely to have body piercings if you both:
- have tattoos
- like heavy metal

NO CORRELATION
People with body piercings are almost exactly as likely as the average person to:
- think it should be legal for two siblings to marry if they intend to not have children
- always tip a server, even when the service is poor
- like string cheese

Bowling Skills

52% of people aren't easily startled.

But among those who are good at bowling, **66%** aren't easily startled.

28% of people say their family considers them to be a black sheep.

But among those who are good at bowling, only **15%** say their family considers them to be a black sheep.

46% of people are good tree climbers.

But among those who are good at bowling, **57%** are good tree climbers.

DOUBLE PLAY

You're extremely likely to be good at bowling if you both:

- are a good Frisbee thrower
- don't have to have the last word in an argument

NO CORRELATION

People who are good at bowling are almost exactly as likely as the average person to:

- like the smell of gasoline
- be bad dancers
- prefer giving family members funny greeting cards, rather than serious ones

Boycotting

32% of people say they have a lot of Jewish friends.

But among those who have never boycotted a company, only **18%** say they have a lot of Jewish friends.

46% of people are willing to pay more for organic foods.

But among those who have never boycotted a company, only **32%** are willing to pay more for organic foods.

57% of people say they are more likely to clam up than talk too much when they're around someone they're attracted to.

But among those who have never boycotted a company, **70%** say they are more likely to clam up than talk too much when they're around someone they're attracted to.

DOUBLE PLAY

You're extremely likely to have never boycotted a company if you both:

- don't get stressed about the holidays
- prefer Jake Gyllenhaal's movies to Maggie Gyllenhaal's

NO CORRELATION

People who have never boycotted a company are almost exactly as likely as the average person to:

- understand HTML
- have friends who are mostly of the same sex
- have been thumb-suckers

Braces

45% of people have ridden on a zip line.

But among those who have worn braces, **57%** have ridden on a zip line.

64% of people think New Yorkers are cooler than Los Angelenos.

But among those who have worn braces, **75%** think New Yorkers are cooler than Los Angelenos.

72% of people say they have lighter skin than others of their ethnicity.

But among those who have worn braces, **82%** say they have lighter skin than others of their ethnicity.

DOUBLE PLAY
You're extremely likely to have worn braces if you both:

- like Bert more than Ernie
- think the government should stop conferring marriages, leaving that to private institutions, and issue only civil unions instead

NO CORRELATION
People who have worn braces are almost exactly as likely as the average person to:

- have honked for a "Honk if you . . ." bumper sticker
- prefer multiflavored Slurpees over single-flavored Slurpees
- have attended private school

Breast Implants

45% of people have positive feelings about athletic scholarships.

But among those with a positive opinion about breast implants, **63%** have positive feelings about athletic scholarships.

33% of people think the world would be worse off if people were 10% dumber but 20% kinder.

But among those with a positive opinion about breast implants, **48%** think the world would be worse off if people were 10% dumber but 20% kinder.

24% of people have run a marathon, or are interested in running one.

But among those with a positive opinion about breast implants, **37%** have run a marathon, or are interested in running one.

DOUBLE PLAY

You're extremely likely to have a positive opinion about breast implants if you both:

- consider yourself to have very little shame
- prefer to read white text on a black background more than black text on a white background

NO CORRELATION

People with a positive opinion about breast implants are almost exactly as likely as the average person to:

- own flannel clothing
- have served as a best man or maid of honor
- be up to date on their tetanus shots

Broken Bones

66% of people enjoy role-playing games.

But among those who have broken a bone, only **56%** enjoy role-playing games.

24% of people have posters on their bedroom wall.

But among those who have broken a bone, only **14%** have posters on their bedroom wall.

42% of people carry around a pocketknife.

But among those who have broken a bone, **52%** carry around a pocketknife.

DOUBLE PLAY

You're extremely likely to have broken a bone if you both:

- say you aren't a coward
- are a good Frisbee thrower

NO CORRELATION

People who have broken a bone are almost exactly as likely as the average person to:

- apply antiperspirant at night, rather than in the morning
- regularly participate in sweepstakes and other giveaways
- like having their feet massaged

Bruising

40% of people think Joe Biden is more likely than the Obama girls to one day end up on a reality show.

But among those who bruise easily, **54%** think Joe Biden is more likely than the Obama girls to one day end up on a reality show.

35% of people have read one of William Faulkner's works.

But among those who bruise easily, **47%** have read one of William Faulkner's works.

57% of people describe their temperament as unflappable.

people who prefer full-size umbrellas over compact ones

people in general
52%

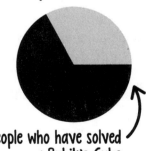

people who have solved a Rubik's Cube
67%

But among those who bruise easily, only **46%** describe their temperament as unflappable.

DOUBLE PLAY

You're extremely likely to bruise easily if you both:

- would drink Tang only under duress
- like to use exclamation points!!!

NO CORRELATION

People who bruise easily are almost exactly as likely as the average person to:

- dislike dishes that are seasoned with rosemary
- have been good sleepers when they were babies
- prefer trashy reality TV over inspirational and up-lifting reality TV

Bumper Stickers

59% of people have square danced.

But among those with bumper stickers on their vehicle, **73%** have square danced.

59% of people own thrift-store clothing.

But among those with bumper stickers on their vehicle, **74%** own thrift-store clothing.

65% of people have eaten venison.

But among those with bumper stickers on their vehicle, **78%** have eaten venison.

You're extremely likely to have bumper stickers on your vehicle if you both:

- have donated to a political campaign
- have a very symmetrical face

NO CORRELATION

People with bumper stickers on their vehicle are almost exactly as likely as the average person to:

- tap their foot to the music
- say their primary home computer is not a laptop
- have tried to raise Sea-Monkeys

Burping at Will

54% of people say that, when entering a chilly pool, they jump in and get it over with rather than wading in little by little.

But among those who can burp at will, **65%** jump in and get it over with rather than wading in little by little.

70% of people like to write lists.

But among those who can burp at will, only **61%** like to write lists.

33% of people have a crown on one or more of their teeth.

But among those who can burp at will, **42%** have a crown on one or more of their teeth.

DOUBLE PLAY

You're extremely likely to be able to burp at will if you both:

- don't get your hair cut at a salon
- like rock candy

NO CORRELATION

People who can burp at will are almost exactly as likely as the average person to:

- have used a fake ID
- have acted dumber than they actually are to improve their social standing
- prefer to have a beautiful obituary over a beautiful gravestone

Bury vs. Flush

47% of people have never gotten a speeding ticket.

But among those who are more likely to bury than flush a dead fish, **61%** have never gotten a speeding ticket.

33% of people say Roger Ebert was their favorite film critic.

But among those who are more likely to bury than flush a dead fish, only **21%** say Roger Ebert was their favorite film critic.

33% of people want their obituary photo to be of "young them," rather than "old them," if they live to be old.

But among those who are more likely to bury than flush a dead fish, **42%** want their obituary photo to be of "young them."

DOUBLE PLAY
You're extremely likely to bury rather than flush a dead fish if you both:

- aren't good at keeping a straight face when you're pulling someone's leg
- have darker skin than others of your ethnicity

NO CORRELATION
People who are more likely to bury than flush a dead fish are almost exactly as likely as the average person to:

- say they would not agree to work a midnight-to-8-a.m. shift for the next 10 years, even if the pay were $1 million a year
- prefer mittens over gloves
- own domain names

Business Cards

47% of people own jazz albums.

But among those who have business cards, **60%** own jazz albums.

57% of people have eaten in a diner at 4 a.m.

But among those who have business cards, **69%** have eaten in a diner at 4 a.m.

46% of people have looked at NSFW stuff at work.

But among those who have business cards, **57%** have looked at NSFW stuff at work.

DOUBLE PLAY
You're extremely likely to have business cards if you both:
- can drive stick shift
- have had your wisdom teeth extracted

NO CORRELATION
People who have business cards are almost exactly as likely as the average person to:
- keep a diary/journal
- have worked in the healthcare industry
- be more likely to sleep on their right side

Camping

50% of people dislike extra chunky pasta sauce.

But among those who don't enjoy camping, **64%** dislike extra chunky pasta sauce.

44% of people aren't good at making paper airplanes.

But among those who don't enjoy camping, **56%** aren't good at making paper airplanes.

39% of people would rather save up for an extravagant vacation than do a series of several more modest mini vacations.

But among those who don't enjoy camping, **50%** would rather save up for an extravagant vacation than do a series of several more modest mini vacations.

DOUBLE PLAY

You're extremely likely to dislike camping if you both:

- have never run a marathon and have no interest in running one
- don't like asparagus

NO CORRELATION

People who don't enjoy camping are almost exactly as likely as the average person to:

- have lost more dignity than money while trying to impress a crush
- prefer to place takeout orders online
- have a positive opinion about breast implants

Cat Lovers

42% of people find it difficult to explain to people what kind of work they do.

But among those who like cats more than dogs, only **31%** find it difficult to explain to people what kind of work they do.

45% of people say that when entering a chilly pool, they wade in little by little rather than jumping in and getting it over with.

But among those who like cats more than dogs, **56%**

wade in little by little rather than jumping in and getting it over with.

71% of people believe it's possible to be a woman born in a man's body, or vice versa.

But among those who like cats more than dogs, **81%** believe it's possible to be a woman born in a man's body, or vice versa.

DOUBLE PLAY
You're extremely likely to like cats more than dogs if you both:
- believe women should be subject to the draft
- don't often pronounce the *T* in often

NO CORRELATION
Cat lovers are almost exactly as likely as the average person to:
- floss regularly
- say the person they are in love with is not in love with them
- prefer their brownies without nuts

Cheating

38% of people have been cheated on.

But among those who have cheated on someone, **69%** have been cheated on.

39% of people have business cards.

But among those who have cheated on someone, **55%** have business cards.

28% of people have dated someone who was totally out of their league.

But among those who have cheated on someone, **39%** have dated someone who was totally out of their league.

DOUBLE PLAY
You're extremely likely to have cheated on someone if you both:
- prefer chemistry to physics
- are good about turning off lights when you leave a room

NO CORRELATION
People who have cheated on someone are almost exactly as likely as the average person to:
- be vegetarians
- not join in the Black Friday shopping frenzy
- say their favorite film is based on a book

Checkers vs. Chess

46% of people have dyed their hair.

But among those who prefer checkers to chess, **60%** have dyed their hair.

51% of people think they would be better off if they weren't so stubborn.

But among those who prefer checkers to chess, **63%** think they would be better off if they weren't so stubborn.

61% of people think they'd make good criminals, if their conscience weren't a factor.

But among those who prefer checkers to chess, only **50%** think they'd make good criminals, if their conscience weren't a factor.

DOUBLE PLAY

You're extremely likely to prefer checkers to chess if you both:

- are not an engineer
- think that fooling kids into thinking that Santa is real is a harmless fib

NO CORRELATION

People who prefer checkers to chess are almost exactly as likely as the average person to:

- think Delaware should cease to be a state and become a suburb of Philadelphia
- say their mother was under 30 when they were born
- dislike yogurt

Chef vs. Maid

60% of people say they could never be in a relationship with a smoker.

But among those who would rather have a personal

chef than a maid, **72%** say they could never be in a relationship with a smoker.

66% of people dislike Nickelback more than Avril Lavigne.

But among those who would rather have a personal chef than a maid, **78%** dislike Nickelback more than Avril Lavigne.

55% of people think that restricting the definition of marriage to only two people discriminates against those in polygamous and polyamorous relationships.

But among those who would rather have a personal chef than a maid, **65%** think that restricting the definition of marriage to only two people discriminates against those in polygamous and polyamorous relationships.

DOUBLE PLAY

You're extremely likely to want a personal chef rather than a maid if you both:

- say your kindergarten teacher would have described you as self-reliant and resilient
- prefer to sip your cold beverages through a straw

NO CORRELATION

People who would rather have a personal chef than a maid are almost exactly as likely as the average person to:

- find freckled faces to be attractive
- be good with kids
- think the mainstream media have a liberal bias

Chicken Pox

39% of people say they have an above-average libido.

But among those who haven't had chicken pox, **69%** say they have an above-average libido.

32% of people dislike cauliflower.

But among those who haven't had chicken pox, **58%** dislike cauliflower.

30% of people say the person they are in love with is not in love with them.

But among those who haven't had chicken pox, **50%** say the person they are in love with is not in love with them.

people who like soggy cereal

people in general
41%

people who pull the tabs off their soda cans
54%

You're extremely likely to not have had chicken pox if you both:

- tend to micromanage when you're in a leadership role
- have never witnessed a meteor shower

NO CORRELATION

People who haven't had chicken pox are almost exactly as likely as the average person to:

- be good at haggling
- say they'd rather have a smiley face than an interrobang as a designated key on their keyboard
- have a hard time with *who* vs. *whom*

Chiropractic Patients

43% of people think a loud baby is more tolerable than a loud dog.

But among those who have visited a chiropractor, **57%** think a loud baby is more tolerable than a loud dog.

57% of people would rather be friends with Prince William than with Prince Harry.

But among those who have visited a chiropractor, **69%** would rather be friends with Prince William than with Prince Harry.

73% of people are not fastidious about locking the bathroom door.

But among those who have visited a chiropractor, **85%** are not fastidious about locking the bathroom door.

DOUBLE PLAY
You're extremely likely to have visited a chiropractor if you both:
- have seen a therapist
- have been under general anesthesia

NO CORRELATION
People who have visited a chiropractor are almost exactly as likely as the average person to:
- like marshmallows in their hot chocolate
- have had asthma
- be able to burp at will

Classical Music

31% of people think they can hold their breath longer than the average person.

But among those who don't like classical music, **49%** think they can hold their breath longer than the average person.

61% of people prefer a combination lock, rather than a lock with a key, for their locker.

But among those who don't like classical music, only **43%** prefer a combination lock for their locker.

62% of people say they are very good at most things that interest them.

But among those who don't like classical music, **77%** say they are very good at most things that interest them.

DOUBLE PLAY

You're extremely likely to dislike classical music if you both:

- don't like V8 juice drinks
- don't own any leather clothing, aside from footwear or belts

NO CORRELATION

People who don't like classical music are almost exactly as likely as the average person to:

- say their emotions influence their decisions more than logic/reasoning
- say they would eat a tasty-looking burger made from flesh cloned from their own body
- think it's better to propose when no one but the two of you is around

Cola Discernment

21% of people say they have never been bitten by an animal.

But among those who can't tell the difference between Coke and Pepsi, **39%** say they have never been bitten by an animal.

56% of people dislike the smell of gasoline.

But among those who can't tell the difference between Coke and Pepsi, **71%** dislike the smell of gasoline.

44% of people regularly skip breakfast.

But among those who can't tell the difference between Coke and Pepsi, only **32%** regularly skip breakfast.

DOUBLE PLAY

You're extremely likely to be unable to tell the difference between Coke and Pepsi if you both:

- don't have a TV in your bedroom
- prefer tea to coffee

NO CORRELATION

People who can't tell the difference between Coke and Pepsi are almost exactly as likely as the average person to:

- have served on a jury
- regularly participate in sweepstakes and other giveaways
- value liberty over security

Cold Weather

37% of people have used the word *obsequious* in conversation.

But among those who prefer cold weather over hot, **48%** have used the word *obsequious* in conversation.

49% of people are Coldplay fans.

But among those who prefer cold weather over hot, only **39%** are Coldplay fans.

37% of people don't make wishes before they blow out their birthday candles.

But among those who prefer cold weather over hot, **47%** don't make wishes before they blow out their birthday candles.

DOUBLE PLAY

You're extremely likely to prefer cold weather over hot if you both:

- don't regularly exercise
- refuse to do the Electric Slide at wedding receptions

NO CORRELATION

People who prefer cold weather over hot are almost exactly as likely as the average person to:

- keep their friends close and their enemies closer
- regularly skip breakfast
- prefer buffalo sauce over barbecue sauce

Cologne/Perfume

56% of people think that restricting the definition of marriage to only two people discriminates against those in polygamous and polyamorous relationships.

But among those who regularly wear cologne/perfume, only **34%** think that restricting the definition of marriage to only two people discriminates against those in polygamous and polyamorous relationships.

19% of people regularly wear a necklace.

But among those who regularly wear cologne/perfume, **39%** regularly wear a necklace.

41% of people prefer amphibians to reptiles.

But among those who regularly wear cologne/perfume, **59%** prefer amphibians to reptiles.

DOUBLE PLAY

You're extremely likely to regularly wear cologne/perfume if you both:

- use air freshener in your home
- dislike lamb

NO CORRELATION

People who regularly wear cologne/perfume are almost exactly as likely as the average person to:

- have a library card
- prefer sweet potatoes to potatoes
- prefer sweet snacks over salty

Corn on the Cob

34% of people prefer pistachios already shelled.

But among those who prefer their corn off the cob, **47%** prefer pistachios already shelled.

67% of people never chew on their pens or pencils.

But among those who prefer their corn off the cob, **78%** never chew on their pens or pencils.

73% of people say they are introverts.

But among those who prefer their corn off the cob, **83%** say they are introverts.

DOUBLE PLAY

You're extremely likely to prefer your corn off the cob if you both:

- prefer peas sans pod
- don't like onions on your burgers

NO CORRELATION

People who prefer their corn off the cob are almost exactly as likely as the average person to:

- be able to recognize the quadratic formula if they saw it
- not regularly wear a necklace
- dislike electric toothbrushes

STATISTICS 101: SELECTION BIAS

When you're polling a group of people, and you want their opinions to be a good representation of the population at large, your poll needs to include a representative sample of people.

Let's suppose that your likelihood of enjoying heavy metal music diminishes with age. If you were to poll only young people or only senior citizens about whether they like heavy metal, you might find that their answers are not representative of the total population. In these cases, we say that selection bias has skewed your results.

Or let's suppose that you want to conduct a poll about privacy issues. You decide to randomly call 1,000 phone numbers listed in the phone book. This will exclude people with unlisted numbers, who may have strong opinions about privacy issues; in addition, depending on the phone book, it may also exclude the growing number of people who don't have a landline telephone. Again, selection bias has skewed your results.

In a certain sense, selection bias is unavoidable when you're conducting polls, because the percentage of the population that refuses to respond to poll questions will always be missing from your sample.

But the effects of selection bias can be minimized by making the effort to understand the demographics of the population whose views you are trying to gauge and making sure that the demographics of the people you poll match them as closely as possible.

Country Music

47% of people like the singer Meat Loaf.

But among those who like country music, **61%** like the singer Meat Loaf.

45% of people have positive feelings about athletic scholarships.

But among those who like country music, **56%** have positive feelings about athletic scholarships.

34% of people struggle more with pride than with low self-esteem.

But among those who like country music, **45%** struggle more with pride than with low self-esteem.

DOUBLE PLAY

You're extremely likely to be a country music fan if you both:

- consider yourself to be patriotic
- have worse-than-average reflexes

NO CORRELATION

People who like country music are almost exactly as likely as the average person to:

- prefer tropical fruits to tropical climates
- be open to being a foster parent
- rather earn more than work less

Court Testimony

36% of people have been on the radio.

But among those who have testified in court, **62%** have been on the radio.

32% of people say Roger Ebert was their favorite film critic.

But among those who have testified in court, **55%** say Roger Ebert was their favorite film critic.

43% of people would drink Tang only under duress.

But among those who have testified in court, **64%** would drink Tang only under duress.

DOUBLE PLAY

You're extremely likely to have testified in court if you both:

- have children
- have ridden in an emergency vehicle

NO CORRELATION

People who have testified in court are almost exactly as likely as the average person to:

- be more likely to sneeze into their sleeve than their hand
- prefer essay test questions over multiple choice questions
- prefer flat-front pants

Credit Cards

69% of people are bad at writing love letters.

But among those who prefer credit cards over debit cards, **85%** are bad at writing love letters.

52% of people are content with their present financial situation.

But among those who prefer credit cards over debit cards, **65%** are content with their present financial situation.

46% of people think clowns are fun.

But among those who prefer credit cards over debit cards, **59%** think clowns are fun.

DOUBLE PLAY

You're extremely likely to prefer credit cards over debit cards if you both:

- own jazz albums
- think that graffiti, no matter how well done, is still a crime

NO CORRELATION

People who prefer credit cards over debit cards are almost exactly as likely as the average person to:

- have fainted in public
- say they would rather change their marital status for a day than their sex
- think it's OK to talk on the phone while on the toilet

Credit Union Members

63% of people have been prescribed a powerful painkiller.

But among credit union members, **78%** have been prescribed a powerful painkiller.

59% of people do their own taxes.

But among credit union members, **73%** do their own taxes.

45% of people prefer a vertical tire swing over a horizontal tire swing.

But among credit union members, **58%** prefer a vertical tire swing.

people who have boycotted a company

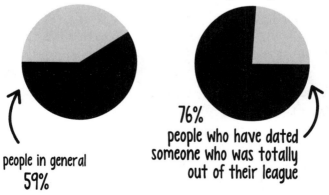

people in general
59%

76%
people who have dated
someone who was totally
out of their league

DOUBLE PLAY

You're extremely likely to be a credit union member if you both:

- are fluent in only one language
- have been involved in scouting

NO CORRELATION

Credit union members are almost exactly as likely as the average person to:

- prefer banana-flavored Runts over bananas
- like Robin Williams better in his serious roles than in his funny roles
- spell well

Cremation Hesitation

36% of people think couples should use traditional wedding vows.

But among those who don't want to be cremated, **48%** think couples should use traditional wedding vows.

51% of people think Mariah Carey has more talent than Christina Aguilera.

But among those who don't want to be cremated, **62%** think Mariah Carey has more talent than Christina Aguilera.

37% of people think the mainstream media have a liberal bias.

But among those who don't want to be cremated, **48%** think the mainstream media have a liberal bias.

DOUBLE PLAY
You're extremely likely to not want to be cremated if you both:

- oppose gay marriage
- would rather host the Academy Awards than the Grammys

NO CORRELATION
People who don't want to be cremated are almost exactly as likely as the average person to:

- have food allergies
- think IKEA is cool
- be fans of Jim Carrey

Cruises

47% of people say their closest platonic friend is not more attractive than them.

But among those who have been on a cruise, **63%** say their closest platonic friend is not more attractive than them.

32% of people have been thrown a surprise party.

But among those who have been on a cruise, **44%** have been thrown a surprise party.

75% of people aren't picky about foods touching each other on their dinner plate.

But among those who have been on a cruise, **88%** aren't picky about foods touching each other on their dinner plate.

DOUBLE PLAY
You're extremely likely to have been on a cruise if you both:

- can swim freestyle
- think fireworks displays are better when they're set to music

NO CORRELATION
People who have been on a cruise are almost exactly as likely as the average person to:

- be more likely to curse than cry when they're hurt
- like tomato soup
- prefer silence when they sleep

Crying and Cursing

40% of people have been in a fistfight.

But among those who are more likely to cry than curse when they're hurt, only **16%** have been in a fistfight.

66% of people think nails that grew twice as quickly would be more annoying than hair that grew twice as quickly.

But among those who are more likely to cry than curse when they're hurt, **86%** think nails that grew twice as quickly would be more annoying.

36% of people prefer cherry cola to regular cola.

But among those who are more likely to cry than curse when they're hurt, **54%** prefer cherry cola to regular cola.

DOUBLE PLAY

You're extremely likely to cry rather than curse when you're hurt if you both:

- are a woman
- have never projectile vomited

NO CORRELATION

People who are more likely to cry than curse when they're hurt are almost exactly as likely as the average person to:

- follow the "if it's yellow, let it mellow" rule of water conservation
- have taken an Internet hiatus
- frequently text

Cursive

30% of people say they have nice handwriting.

But among those who write primarily in cursive, **51%** say they have nice handwriting.

70% of people want whipped cream on top of their sundae.

But among those who write primarily in cursive, only **57%** want whipped cream on top of their sundae.

47% of people would rather have a colostomy bag than a feeding tube.

But among those who write primarily in cursive, **60%** would rather have a colostomy bag than a feeding tube.

DOUBLE PLAY

You're extremely likely to write primarily in cursive if you both:

- have had a urinary tract infection
- have cried over the death of someone you've never met

NO CORRELATION

People who write primarily in cursive are almost exactly as likely as the average person to:

- consider themselves unlucky
- know which directions "port" and "starboard" refer to
- like their shower temperature to be just short of scalding

Death Penalty

58% of people dislike Big Government more than Big Business.

But among those who support capital punishment, only **40%** dislike Big Government more than Big Business.

17% of people think women have it easier than men.

But among those who support capital punishment, **33%** think women have it easier than men.

59% of people find straight hair more attractive than curly hair.

But among those who support capital punishment, **72%** find straight hair more attractive than curly hair.

DOUBLE PLAY

You're extremely likely to support capital punishment if you both:

- think ambition influences success more than privilege
- enjoy fiction more than nonfiction

NO CORRELATION

People who support capital punishment are almost exactly as likely as the average person to:

- belong to a gym or fitness center
- prefer green apples to red apples
- dislike sideburns

Decision Making

58% of people don't have sinus issues.

But among those whose emotions influence their decisions more than logic or reasoning, **77%** don't have sinus issues.

53% of people say they have average (or worse) reflexes.

But among those whose emotions influence their decisions more than logic or reasoning, **71%** say they have average (or worse) reflexes.

22% of people would rather see increased funding to the arts than to the sciences.

But among those whose emotions influence their decisions more than logic or reasoning, **38%** would rather see increased funding to the arts than to the sciences.

DOUBLE PLAY

You're extremely likely to say that your emotions influence your decisions more than logic or reasoning if you both:

- are a woman
- leave your emails in your inbox after you've read them

NO CORRELATION

People whose emotions influence their decisions more than logic or reasoning are almost exactly as likely as the average person to:

- prefer cushioned toilet seats
- prefer compact umbrellas over full-size umbrellas
- think it's sometimes OK for a significant other to keep photos of old flames

Deep-Dish Pizza

60% of people prefer mechanical pencils.

But among those who prefer deep-dish pizza to thin-crust pizza, **71%** prefer mechanical pencils.

60% of people don't own an iPad or other tablet.

But among those who prefer deep-dish pizza to thin-crust pizza, **70%** don't own an iPad or other tablet.

49% of people prefer creamy salad dressings.

But among those who prefer deep-dish pizza to thin-crust pizza, **58%** prefer creamy salad dressings.

DOUBLE PLAY

You're extremely likely to prefer deep-dish pizza to thin-crust pizza if you both:

- pray
- would like to take a martial arts class

NO CORRELATION

People who prefer deep-dish pizza to thin-crust pizza are almost exactly as likely as the average person to:

- have idiosyncratic rules about how they eat their Kit-Kat bar
- have a throwaway email address
- be married

Diaries

47% of people bring their lunch more often than they buy their lunch.

But among those who keep a diary or journal, **65%** bring their lunch more often than they buy their lunch.

27% of people make New Year's resolutions.

But among those who keep a diary or journal, **45%** make New Year's resolutions.

43% of people are involved in volunteer work.

But among those who keep a diary or journal, **58%** are involved in volunteer work.

DOUBLE PLAY

You're extremely likely to keep a diary or journal if you both:

- like to write lists
- have read *The Hunger Games*

NO CORRELATION

People who keep a diary or journal are almost exactly as likely as the average person to:

- think Adam and Eve had belly buttons
- prefer soft-shell tacos
- say they're typically believed to be younger than they are

Dodge Ball vs. Kickball

69% of people prefer physics to chemistry.

But among those who prefer dodge ball over kickball, **81%** prefer physics to chemistry.

59% of people find straight hair more attractive than curly hair.

But among those who prefer dodge ball over kickball, **68%** find straight hair more attractive than curly hair.

47% of people prefer cold weather over hot.

But among those who prefer dodge ball over kickball, **56%** prefer cold weather over hot.

DOUBLE PLAY
You're extremely likely to prefer dodge ball over kickball if you both:
- enjoy role-playing games
- don't bruise easily

NO CORRELATION
People who prefer dodge ball over kickball are almost exactly as likely as the average person to:
- live alone
- smoke
- believe aliens exist

Donating Blood

49% of people say they feel like grown-ups.

But among those who have donated blood, **60%** say they feel like grown-ups.

- - - - - - - - - - - - - - -

28% of people like creamed chipped beef.

But among those who have donated blood, **38%** like creamed chipped beef.

- - - - - - - - - - - - - - -

64% of people say they are well informed about world events.

But among those who have donated blood, **73%** say they are well informed about world events.

DOUBLE PLAY

You're extremely likely to have donated blood if you both:

- would agree to staff a kissing booth to raise funds for your favorite charity
- are a registered organ donor

NO CORRELATION

People who have donated blood are almost exactly as likely as the average person to:

- be in better shape now than five years ago
- like horror movies
- have cried during *Titanic*

Double-Jointedness

66% of people like Liv Tyler more than Steven Tyler.

But among those who are double-jointed, **85%** like Liv Tyler more than Steven Tyler.

53% of people say their signatures are illegible.

But among those who are double-jointed, **72%** say their signatures are illegible.

62% of people are fans of Penn & Teller.

But among those who are double-jointed, **80%** are fans of Penn & Teller.

DOUBLE PLAY

You're extremely likely to be double-jointed if you both:

- have had pink eye
- watch the Super Bowl for the commercials rather than the game

NO CORRELATION

People who are double-jointed are almost exactly as likely as the average person to:

- say they'd rather have a maid than a personal chef
- prefer their toilet paper to hang against the wall
- not be willing to let a stranger borrow a cellphone to make a call

people who like marshmallows in their hot chocolate

people in general
69%

people who prefer sprinkles
on their ice cream cones
86%

Emergency Vehicles

50% of people have ridden a motorcycle.

But among those who have ridden in an emergency vehicle, **62%** have ridden a motorcycle.

38% of people have a higher opinion of Apple's iOS than of Google's Android OS.

But among those who have ridden in an emergency vehicle, **50%** have a higher opinion of Apple's iOS than of Google's Android OS.

35% of people have mistakenly used the public restroom of the opposite sex.

But among those who have ridden in an emergency vehicle, **46%** have mistakenly used the public restroom of the opposite sex.

DOUBLE PLAY

You're extremely likely to have ridden in an emergency vehicle if you both:

- have donated blood
- like the singer Meat Loaf

NO CORRELATION

People who have ridden in an emergency vehicle are almost exactly as likely as the average person to:

- say they would rather repeal the Nineteenth Amendment, which gave women the right to vote, than the Twenty-First Amendment, which ended the alcohol prohibition
- like licorice
- say a physical feature is the thing they like least about themselves

Exclamation Points

67% of people find Australians sexy.

But among those who don't like to use exclamation points, only **55%** find Australians sexy.

44% of people would rather be friends with Prince Harry than Prince William.

But among those who don't like to use exclamation

points, only **32%** would rather be friends with Prince Harry than Prince William.

74% of people are more likely to arrange a bookshelf based on subject than book size or color.

But among those who don't like to use exclamation points, **85%** are more likely to arrange a bookshelf based on subject than book size or color.

DOUBLE PLAY
You're extremely likely to not like using exclamation points if you both:
- are an overachiever
- are unaware of having ever been rickrolled

NO CORRELATION
People who don't like to use exclamation points are almost exactly as likely as the average person to:
- own raincoats
- floss regularly
- prefer iced tea to lemonade

Exercise Habits

64% of people think New Yorkers are cooler than Los Angelenos.

But among those who don't regularly exercise, **77%** think New Yorkers are cooler than Los Angelenos.

60% of people stay in the movie theater until the credits end.

But among those who don't regularly exercise, **70%** stay in the movie theater until the credits end.

28% of people like creamed chipped beef.

But among those who don't regularly exercise, **37%** like creamed chipped beef.

DOUBLE PLAY

You're extremely likely to not regularly exercise if you both:

- aren't an overachiever
- can't name more than one song by Flo Rida

NO CORRELATION

People who don't regularly exercise are almost exactly as likely as the average person to:

- regularly check their horoscope
- use the same password for pretty much everything
- buy the extended warranty on big-ticket items

Extroverts

35% of people say mosquitos prefer biting them more than the average person.

But among extroverts, **56%** say mosquitos prefer biting them more than the average person.

30% of people are Justin Timberlake fans.

But among extroverts, **49%** are Justin Timberlake fans.

41% of people aren't bothered by parents who kiss their adult children on the lips.

But among extroverts, **60%** aren't bothered by parents who kiss their adult children on the lips.

DOUBLE PLAY

You're extremely likely to be an extrovert if you both:
- consider yourself very physically affectionate
- like anchovies

NO CORRELATION

Extroverts are almost exactly as likely as the average person to:
- pig out when they're upset
- have dimples
- take off their shoes when they enter their home

Eyesight

38% of people say they have an above-average libido.

But among those who don't wear glasses or contacts, **51%** say they have an above-average libido.

56% of people always wash their hands after using a public restroom.

But among those who don't wear glasses or contacts, only **43%** always wash their hands after using a public restroom.

36% of people say that as children, they liked when things were neat and tidy.

But among those who don't wear glasses or contacts, **48%** say that as children, they liked when things were neat and tidy.

DOUBLE PLAY

You're extremely likely to not wear glasses or contacts if you both:

- don't have to have the last word in an argument
- never spill things on your keyboard

NO CORRELATION

People who don't wear glasses or contacts are almost exactly as likely as the average person to:

- prefer emoticons with noses
- like the smell of citronella
- have a TV in their bedroom

STATISTICS 101: CONFOUNDING VARIABLES

If you were to create a chart that depicts ice cream sales and drowning incidents over time, you would find that they both rise and fall at about the same times of the year.

Does that mean that higher ice cream sales cause drownings or vice versa?

No. Statisticians call this a "spurious relationship" because neither causes the other.

The most obvious explanation for the correlation is that a confounding variable exists. Confounding variables, also called hidden variables, are pertinent factors that we may not necessarily take into account when we observe a connection between two unrelated things.

In this simple example, the weather is the confounding variable. People are more likely to eat ice cream when the weather is warm, and they're also more likely to swim (and drown) when it's warm out.

Fainting

32% of people say they have a lot of Jewish friends.

But among those who have fainted in public, **49%** say they have a lot of Jewish friends.

54% of people are wine drinkers.

But among those who have fainted in public, **68%** are wine drinkers.

44% of people think Saturday night is a better night to party than Friday night.

But among those who have fainted in public, **56%** think Saturday night is a better night to party than Friday night.

DOUBLE PLAY

You're extremely likely to have fainted in public if you both:

- have dyed your hair
- regularly drink alcohol

NO CORRELATION

People who have fainted in public are almost exactly as likely as the average person to:

- have visited a chiropractor
- prefer dark toast to light toast
- have dandruff

Fan Mail

75% of people find it hard to tell when someone's just being flirty, versus when they're really interested.

But among those who have written a fan letter to a celebrity, only **58%** find it hard to tell when someone's just being flirty, versus when they're really interested.

26% of people like wall-to-wall carpeting more than hardwood floors.

But among those who have written a fan letter to a celebrity, **37%** like wall-to-wall carpeting more than hardwood floors.

21% of people have dated someone they met on a dating website.

But among those who have written a fan letter to a celebrity, **32%** have dated someone they met on a dating website.

DOUBLE PLAY
You're extremely likely to have written a fan letter to a celebrity if you both:

- prefer Good & Plenty over Mike and Ike candies
- would rather time-travel to the past than to the future

NO CORRELATION
People who have written a fan letter to a celebrity are almost exactly as likely as the average person to:

- expect Social Security to be insolvent when they retire
- prefer to visit Philly for the Liberty Bell than for the cheesesteaks
- like cantaloupe

Filtered Water

40% of people think they would recognize the Dalai Lama if he walked past them in jeans and a Lakers jersey.

But among those who filter all their tap water, **57%** think they would recognize the Dalai Lama if he walked past them in jeans and a Lakers jersey.

69% of people are bad at writing love letters.

But among those who filter all their tap water, **82%** are bad at writing love letters.

49% of people say they are more focused on the future than the present.

But among those who filter all their tap water, **60%** say they are more focused on the future than the present.

DOUBLE PLAY

You're extremely likely to filter all your tap water if you both:

- wash new clothes before you wear them
- like cauliflower

NO CORRELATION

People who filter all their tap water are almost exactly as likely as the average person to:

- like professional wrestling
- have a best friend
- have bumper stickers on their vehicle

Fistfights

67% of people know how to tie a necktie.

But among those who have been in a fistfight, **82%** know how to tie a necktie.

42% of people carry around a pocketknife.

But among those who have been in a fistfight, **57%** carry around a pocketknife.

43% of people have projectile vomited.

But among those who have been in a fistfight, **56%** have projectile vomited.

DOUBLE PLAY
You're extremely likely to have been in a fistfight if you both:
- don't wear pajamas to bed
- own domain names

NO CORRELATION
People who have been in a fistfight are almost exactly as likely as the average person to:
- find it hard to resist picking at scabs
- have a strong preference for a particular Monopoly game piece
- prefer flavored water over unflavored water

Flag Burning

43% of people dislike the scent of menthol.

But among those who think flag burning should be illegal, **58%** dislike the scent of menthol.

38% of people don't have any sisters.

But among those who think flag burning should be illegal, only **24%** don't have any sisters.

26% of people wouldn't want to know the sex of a baby before it's born.

But among those who think flag burning should be

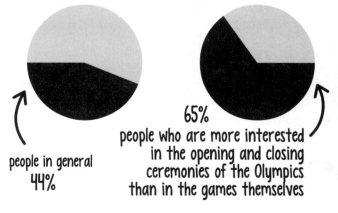

people who feel more confident in dressy clothes than comfortable clothes

people in general
44%

65%
people who are more interested in the opening and closing ceremonies of the Olympics than in the games themselves

illegal, **37%** wouldn't want to know the sex of a baby before it's born.

DOUBLE PLAY
You're extremely likely to think that flag burning should be illegal if you both:
- consider voting a moral duty
- value security over liberty

NO CORRELATION
People who think flag burning should be illegal are almost exactly as likely as the average person to:
- prefer to have turn-by-turn directions written out, rather than following a map
- prefer thin-crust pizza to deep-dish pizza
- prefer carbonated beverages

Flossing

56% of people don't have a strong preference for a particular Monopoly game piece.

But among those who floss regularly, **71%** don't have a strong preference for a particular Monopoly game piece.

27% of people regularly use mouthwash.

But among those who floss regularly, **43%** regularly use mouthwash.

40% of people have their best friend's phone number memorized.

But among those who floss regularly, **53%** have their best friend's phone number memorized.

DOUBLE PLAY
You're extremely likely to floss regularly if you both:
- prefer earbuds to headphones
- like beets

NO CORRELATION
People who floss regularly are almost exactly as likely as the average person to:
- call cola "pop" rather than "soda"
- like curly fries
- say their left leg is on the bottom when they sit cross-legged

Foreign Accents

69% of people can roll their Rs.

But among those who are good at imitating foreign accents, **84%** can roll their Rs.

67% of people are interested in geology.

But among those who are good at imitating foreign accents, **79%** are interested in geology.

60% of people think they'd make good criminals, if their conscience weren't a factor.

But among those who are good at imitating foreign

accents, **71%** think they'd make good criminals, if their conscience weren't a factor.

DOUBLE PLAY
You're extremely likely to be good at imitating foreign accents if you both:
- are comfortable speaking in public
- would rather compete on *Jeopardy!* than *Wheel of Fortune*

NO CORRELATION
People who are good at imitating foreign accents are almost exactly as likely as the average person to:
- have nonsummer birthdays
- filter all their tap water
- prefer the term *couch* to *sofa*

Fraternities and Sororities

63% of people like Jerry Seinfeld.

But among those with a positive opinion about fraternities and sororities, **75%** like Jerry Seinfeld.

22% of people like super-cheesy pickup lines.

But among those with a positive opinion about fraternities and sororities, **35%** like super-cheesy pickup lines.

57% of people would prefer to work at a place where they are considered a hotshot and are doted on, rather than a place where they're in the middle of the pack.

But among those with a positive opinion about fraternities and sororities, **69%** would prefer to work at a place where they are considered a hotshot and are doted on, rather than a place where they're in the middle of the pack.

DOUBLE PLAY
You're extremely likely to have a positive opinion about fraternities and sororities if you both:

- think Halloween is more fun when you're of legal drinking age than when you're a kid
- don't suffer from canker sores

NO CORRELATION
People with a positive opinion about fraternities and sororities are almost exactly as likely as the average person to:

- eat corn on the cob in a spiral
- agree that it's better to have loved and lost than never to have loved at all
- like V8 juice drinks

Glow Bowling

61% of people like Dr Pepper.

But among those who have gone glow bowling, **78%** like Dr Pepper.

29% of people want a tattoo.

But among those who have gone glow bowling, **42%** want a tattoo.

43% of people think Saturday night is a better night to party than Friday night.

But among those who have gone glow bowling, **56%** think Saturday night is a better night to party than Friday night.

DOUBLE PLAY
You're extremely likely to have gone glow bowling if you both:
- went to a high school that had a large marching band program
- dine out frequently

NO CORRELATION
People who have gone glow bowling are almost exactly as likely as the average person to:
- say they are more scared of bees than the average person
- watch the Super Bowl for the commercials, rather than for the game
- filter all their tap water

Going Dutch

31% of people don't believe it's their obligation to buy their designated driver's soft drinks.

But among those who think the costs should be split

during a date, **49%** don't believe it's their obligation to buy their designated driver's soft drinks.

41% of people are married.

But among those who think the costs should be split during a date, only **25%** are married.

19% of people would welcome our new insect overlords, rather than fight to the death, if giant mutated ants attempted to destroy humanity.

But among those who think the costs should be split during a date, **32%** would welcome our new insect overlords.

DOUBLE PLAY

You're extremely likely to think the costs should be split during a date if you both:

- were not born in the United States
- ignore QR codes

NO CORRELATION

People who think the costs should be split during a date are almost exactly as likely as the average person to:

- have older siblings
- agree that it's better to have loved and lost than never to have loved at all
- prefer lemons over limes

Good Dancers

55% of people are good at giving massages.

But among those who say they're good dancers, **76%** are good at giving massages.

34% of people regularly use ChapStick or other lip balm.

But among those who say they're good dancers, **51%** regularly use ChapStick or other lip balm.

32% of people have been thrown a surprise party.

But among those who say they're good dancers, **48%** have been thrown a surprise party.

DOUBLE PLAY

You're extremely likely to be a good dancer if you both:

- find it easy to tell when someone's just being flirty, versus when they're really interested
- don't put a cap on uppercase *J*s

NO CORRELATION

Good dancers are almost exactly as likely as the average person to:

- have a negative opinion about surrogate pregnancy
- have never pulled an all-nighter
- have written a fan letter to a celebrity

Good Singers

33% of people are good at imitating foreign accents.

But among those who are good singers, **48%** are good at imitating foreign accents.

35% of people say they regularly pray.

But among those who are good singers, **49%** say they regularly pray.

57% of people think that if they could redo high school, knowing everything they know now, they would be just as uncool as they were the first time.

But among those who are good singers, **69%** think that if they could redo high school, knowing everything they know now, they would be just as uncool as they were the first time.

DOUBLE PLAY

You're extremely likely to be a good singer if you both:

- are able to read music
- have a higher opinion of Apple's iOS than of Google's Android OS

NO CORRELATION

People who are good singers are almost exactly as likely as the average person to:

- dislike honey-mustard sauce
- never pee in the shower
- prefer foreign cars

Grapes

60% of people prefer red apples to green apples.

But among those who prefer purple grapes to green grapes, **74%** prefer red apples.

52% of people are Foo Fighters fans.

But among those who prefer purple grapes to green grapes, only **40%** are Foo Fighters fans.

34% of people say they struggle more with pride than with low self-esteem.

But among those who prefer purple grapes to green grapes, **43%** say they struggle more with pride than with low self-esteem.

DOUBLE PLAY

You're extremely likely to prefer purple grapes to green grapes if you both:

- say mosquitos prefer biting you more than the average person
- prefer a hot breakfast over a cold breakfast

NO CORRELATION

People who prefer purple grapes to green grapes are almost exactly as likely as the average person to:

- like wall-to-wall carpeting more than hardwood floors
- believe aliens exist
- have strong feelings about whether physician-assisted suicide should be legal

Greeting Cards

74% of people prefer Alex P. Keaton of *Family Ties* to Mike Seaver of *Growing Pains*.

But among those who tend to give heartfelt greeting cards, rather than funny ones, to family members, **92%** prefer Alex P. Keaton of *Family Ties* to Mike Seaver of *Growing Pains*.

38% of people think Holocaust denial should be illegal.

But among those who tend to give heartfelt greeting cards, rather than funny ones, to family members, **54%** think Holocaust denial should be illegal.

35% of people think Detroit would win in a fight against Chicago.

But among those who tend to give heartfelt greeting cards, rather than funny ones, to family members, **49%** think Detroit would win in a fight against Chicago.

DOUBLE PLAY

You're extremely likely to give heartfelt greeting cards, rather than funny ones, to family members if you both:

- think nails that grow twice as quickly as average would be more annoying than hair that grows twice as quickly as average
- prefer Home Depot over Lowe's

People who tend to give heartfelt greeting cards to family members are almost exactly as likely as the average person to:

- have a best friend
- follow the "if it's yellow, let it mellow" rule of water conservation
- like asparagus

Guacamole

50% of people would bare it all on a nude beach.

But among those who don't like guacamole, only **38%** would bare it all on a nude beach.

14% of people prefer banana-flavored Runts over bananas.

But among those who don't like guacamole, **26%** prefer banana-flavored Runts over bananas.

28% of people tend to stir their drinks counterclockwise.

But among those who don't like guacamole, **39%** tend to stir their drinks counterclockwise.

DOUBLE PLAY
You're extremely likely to dislike guacamole if you both:

- don't like sushi
- aren't interested in going into business for yourself

people who are members of warehouse clubs

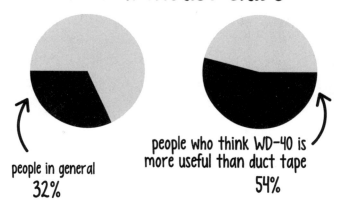

people in general
32%

people who think WD-40 is more useful than duct tape
54%

NO CORRELATION

People who don't like guacamole are almost exactly as likely as the average person to:

- be unable to type without looking at the keyboard
- be able to double-dutch
- prefer green grapes to purple grapes

Hair Dyeing

53% of people are interested in sign language.

But among those who have dyed their hair, **63%** are interested in sign language.

60% of people would rather play a bass drum than a tuba in a marching band.

But among those who have dyed their hair, **69%** people would rather play a bass drum than a tuba in a marching band.

39% of people are good at remembering birthdays and anniversaries.

But among those who have dyed their hair, **49%** are good at remembering birthdays and anniversaries.

DOUBLE PLAY
You're extremely likely to have dyed your hair if you both:
- know how to use a sewing machine
- have tattoos

NO CORRELATION
People who have dyed their hair are almost exactly as likely as the average person to:
- prefer fame over fortune
- have gone to sleep-away camp
- be more likely to arrange a bookshelf based on book size or color than subject

Handwriting

36% of people doodle frequently.

But among those with nice handwriting, **50%** doodle frequently.

23% of people write in cursive.

But among those with nice handwriting, **37%** write in cursive.

46% of people like anime.

But among those with nice handwriting, only **34%** like anime.

DOUBLE PLAY

You're extremely likely to have nice handwriting if you both:

- don't prefer to use an electric razor
- like to use exclamation points!!!

NO CORRELATION

People with nice handwriting are almost exactly as likely as the average person to:

- regularly use a Laundromat
- have children
- have a good sense of direction

Horoscopes

22% of people are claustrophobic.

But among those who regularly check their horoscope, **48%** are claustrophobic.

50% of people have nice skin.

But among those who regularly check their horoscope, **75%** people have nice skin.

40% of people press the "door close" button after selecting their elevator floor.

But among those who regularly check their horoscope, **65%** press the "door close" button after selecting their elevator floor.

DOUBLE PLAY

You're extremely likely to regularly check your horoscope if you both:

- have never taken a defensive-driving course
- prefer to read white text on a black background more than black text on a white background

NO CORRELATION

People who regularly check their horoscope are almost exactly as likely as the average person to:

- prefer mint gum to non-mint gum
- like Brussels sprouts
- think it's never OK for a significant other to keep photos of old flames

Horror Movies

15% of people can spin a basketball on their finger.

But among those who like horror movies, **30%** can spin a basketball on their finger.

40% of people are fans of *Arrested Development*.

But among those who like horror movies, **52%** are fans of *Arrested Development*.

54% of people root for Wile E. Coyote over the Road Runner.

But among those who like horror movies, **66%** root for Wile E. Coyote over the Road Runner.

DOUBLE PLAY

You're extremely likely to enjoy horror movies if you both:

- want a tattoo
- believe that churches and other places of worship should not be tax exempt

NO CORRELATION

People who like horror movies are almost exactly as likely as the average person to:

- rather have a smiley face than an interrobang as a designated key on their keyboard
- have solved a Rubik's Cube
- have illegible signatures

HTML Comprehension

47% of people know who actress Felicia Day is.

But among those who understand HTML, **59%** know who actress Felicia Day is.

46% of people have looked at NSFW stuff at work.

But among those who understand HTML, **55%** have looked at NSFW stuff at work.

77% of people have never purposely listened to a Justin Bieber song.

But among those who understand HTML, **85%** have never purposely listened to a Justin Bieber song.

DOUBLE PLAY

You're extremely likely to understand HTML if you both:

- regularly back up your files
- are more likely to do electric work on your own than plumbing work

NO CORRELATION

People who understand HTML are almost exactly as likely as the average person to:

- consider themselves less generous than the average person
- be able to differentiate between a tangerine and a clementine
- be able to pop a wheelie

STATISTICS 101:
CORRELATION AND CAUSATION

Just because two things are correlated, it doesn't necessarily mean that there is a causal relationship between the two.

For instance, the rate of autism diagnoses has been rising over the past 15 years, as have sales of organic foods. That doesn't mean that the rise in one causes the rise in the other.

However, some correlations are indeed the result of a causal relationship. For instance, we know, based on empirical observations, that tall parents tend to have tall children, and short parents tend to have short children. The height of the parent is a cause (although not the only determining factor) of the child's height.

How can you tell whether a causal connection exists between two things that appear to be correlated? Ask yourself these questions:

1. How strong is the correlation between A and B? A weak correlation makes a causal relationship unlikely.

2. By what process might A cause B, or B cause A? If you can come up with a plausible process, you're on your way to establishing a causal relationship, but even then, it's possible that your explanation, while plausible, is nonetheless incorrect.

3. Can you think of any confounding variables (see page 68) that might be the root cause of both A and B? If so, that may be the more likely explanation.

4. If A didn't exist, would B continue? If B would indeed continue, then A can't be the exclusive cause of B.

iPods/iPads

32% of people have had to make a life-or-death decision.

But among those who don't own iPods/iPads, only **21%** have had to make a life-or-death decision.

54% of people aren't willing to pay more for organic foods.

But among those who don't own iPods/iPads, **65%** aren't willing to pay more for organic foods.

28% of people tend to stir their drinks counter-clockwise.

But among those who don't own iPods/iPads, **38%** tend to stir their drinks counterclockwise.

DOUBLE PLAY
You're extremely likely to not own an iPod/iPad if you both:
- don't own skinny jeans
- dislike Avril Lavigne more than Nickelback

NO CORRELATION
People who don't own iPods/iPads are almost exactly as likely as the average person to:
- be credit union members
- have served on a jury
- have had wisdom teeth extracted

Jugglers

40% of people prefer Bill Clinton over Hillary Clinton.

But among those who can juggle, **59%** prefer Bill Clinton over Hillary Clinton.

53% of people prefer the top bunk to the bottom bunk.

But among those who can juggle, **70%** prefer the top bunk to the bottom bunk.

64% of people think the penny should be done away with.

But among those who can juggle, **80%** think the penny should be done away with.

DOUBLE PLAY

You're extremely likely to be able to juggle if you both:

- can spin a basketball on your finger
- can hold your breath longer than the average person

NO CORRELATION

People who can juggle are almost exactly as likely as the average person to:

- dislike extra chunky pasta sauce
- like classical music
- prefer to use an electric razor

Jury Duty

37% of people have used the word *obsequious* in conversation.

But among those who have served on a jury, **59%** have used the word *obsequious* in conversation.

41% of people say they have good posture.

But among those who have served on a jury, **62%** say they have good posture.

61% of people are fans of Penn & Teller.

But among those who have served on a jury, **82%** are fans of Penn & Teller.

DOUBLE PLAY
You're extremely likely to have served on a jury if you both:
- have been on a cruise
- prefer gold-colored jewelry

NO CORRELATION
People who have served on a jury are almost exactly as likely as the average person to:
- prefer plain M&M's
- oppose the "cry it out" approach to child sleep issues
- prefer multiple choice test questions over essay questions

LA vs. NYC

40% of people support capital punishment.

But among those who think Los Angelenos are cooler than New Yorkers, **55%** support capital punishment.

32% of people are members of warehouse clubs.

But among those who think Los Angelenos are cooler than New Yorkers, **47%** are members of warehouse clubs.

26% of people would rather change their marital status for a day than their sex.

But among those who think Los Angelenos are cooler than New Yorkers, **39%** would rather change their marital status for a day than their sex.

DOUBLE PLAY

You're extremely likely to think Los Angelenos are cooler than New Yorkers if you both:
- prefer doughnuts to muffins
- are more likely to become vegetarian for health reasons than ethical reasons

NO CORRELATION

People who think Los Angelenos are cooler than New Yorkers are almost exactly as likely as the average person to:
- have had chicken pox
- like a cappella music
- have high cholesterol

Lactose Intolerance

42% of people are interested in going into business for themselves.

But among those who are lactose intolerant, only **14%** are interested in going into business for themselves.

46% of people like beets.

But among those who are lactose intolerant, **68%** like beets.

46% of people would rather ride a waterslide than a roller coaster.

But among those who are lactose intolerant, **64%** would rather ride a waterslide than a rollercoaster.

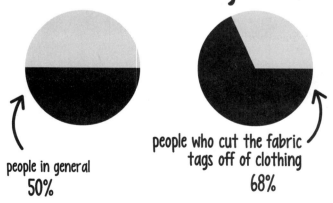

people who have ridden on a motorcycle

people in general
50%

people who cut the fabric tags off of clothing
68%

DOUBLE PLAY

You're extremely likely to be lactose intolerant if you both:

- prefer literature classes over composition classes
- like onions on your burgers

NO CORRELATION

People who are lactose intolerant are almost exactly as likely as the average person to:

- have gone glow bowling
- think prostitution should be legal
- have had athlete's foot

Lady Gaga Fans

72% of people believe it's possible to be a woman born in a man's body or vice versa.

But among Lady Gaga fans, **84%** believe it's possible to be a woman born in a man's body or vice versa.

77% of people have a positive opinion about surrogate pregnancy.

But among Lady Gaga fans, **88%** people have a positive opinion about surrogate pregnancy.

19% of people regularly tweet on Twitter.

But among Lady Gaga fans, **29%** regularly tweet on Twitter.

You're extremely likely to be a Lady Gaga fan if you both:

- supported Barack Obama's reelection bid
- consider yourself to have very little shame

NO CORRELATION

Lady Gaga fans are almost exactly as likely as the average person to:

- have never taken an Internet hiatus
- own a bicycle
- have had an imaginary friend when they were a child

Landlines

33% of people think Google hasn't done a good job of adhering to its unofficial "Don't be evil" motto.

But among those who have a landline phone, **45%** think Google hasn't done a good job of adhering to its unofficial "Don't be evil" motto.

41% of people are fans of *Arrested Development*.

But among those who have a landline phone, only **29%** are fans of *Arrested Development*.

33% of people dislike Avril Lavigne more than Nickelback.

But among those who have a landline phone, **44%** dislike Avril Lavigne more than Nickelback.

DOUBLE PLAY

You're extremely likely to have a landline phone if you both:

- prefer gold-colored jewelry to silver-colored jewelry
- watch TV news

NO CORRELATION

People who have a landline phone are almost exactly as likely as the average person to:

- think World War III will occur in the next 25 years
- be unable to tell the difference between Coke and Pepsi
- like sea scallops

Left Handers

49% of people think that men who wear hats indoors are committing a faux pas.

But among left-handers, **69%** think that men who wear hats indoors are committing a faux pas.

53% of people say their closest platonic friend is more attractive than they are.

But among left-handers, only **36%** say their closest platonic friend is more attractive than they are.

31% of people would prefer to be the world's most talented visual artist, rather than the world's most talented musician.

But among left-handers, **45%** would prefer to be the world's most talented visual artist, rather than the world's most talented musician.

DOUBLE PLAY
You're extremely likely to be left-handed if you both:
- prefer to be photographed from your left side
- don't find it hard to resist picking at scabs

NO CORRELATION
Left-handers are almost exactly as likely as the average person to:
- think sunrise is more beautiful than sunset
- like the smell of cedar
- be willing to bare it all on a nude beach

Lemons vs. Limes

60% of people prefer fiction to nonfiction.

But among those who prefer lemons over limes, **72%** prefer fiction to nonfiction.

71% of people like Robin Williams better in his funny roles than in his serious roles.

But among those who prefer lemons over limes, **82%** like Robin Williams better in his funny roles than in his serious roles.

59% of people prefer Hillary Clinton over Bill Clinton.

But among those who prefer lemons over limes, **68%** prefer Hillary Clinton over Bill Clinton.

DOUBLE PLAY
You're extremely likely to prefer lemons over limes if you both:
- don't like Dr Pepper
- consider yourself more generous than the average person

NO CORRELATION
People who prefer lemons over limes are almost exactly as likely as the average person to:
- have posted their photo on a site where others vote on their attractiveness
- prefer bagels over croissants
- be tactful

Leno vs. Letterman

61% of people prefer doughnuts to muffins.

But among those who prefer Leno to Letterman, **74%** prefer doughnuts to muffins.

51% of people think Mariah Carey has more talent than Christina Aguilera.

But among those who prefer Leno to Letterman, **63%** think Mariah Carey has more talent than Christina Aguilera.

65% of people have brothers.

But among those who prefer Leno to Letterman, **76%** have brothers.

DOUBLE PLAY
You're extremely likely to prefer Leno to Letterman if you both:
- would rather fight 100 duck-size horses than one horse-size duck
- think we're too generous with foreign aid

NO CORRELATION
People who prefer Leno to Letterman are almost exactly as likely as the average person to:
- rather have stinky farts than loud farts
- say they could never be in a relationship with a smoker
- like yogurt

Licorice

58% of people like rock candy.

But among those who like licorice, **71%** like rock candy.

68% of people know the difference between amps, volts, and ohms.

But among those who like licorice, **79%** know the difference between amps, volts, and ohms.

60% of people stay in the movie theater until the credits end.

But among those who like licorice, **69%** stay in the movie theater until the credits end.

DOUBLE PLAY
You're extremely likely to like licorice if you both:
- use shoe polish
- have a generally positive opinion about France

NO CORRELATION
People who like licorice are almost exactly as likely as the average person to:
- watch TV news
- say their families moved around a lot during their childhood
- make New Year's resolutions

Losing One's Appetite

57% of people say they are more likely to clam up than talk too much when they're around someone they're attracted to.

But among those who lose their appetite when they're upset, **69%** say they are more likely to clam up than talk too much when they're around someone they're attracted to.

32% of people say they had a negative first impression of Pope Francis.

But among those who lose their appetite when they're upset, **44%** had a negative first impression of Pope Francis.

38% of people ascend the stairs two at a time.

But among those who lose their appetite when they're upset, **48%** ascend the stairs two at a time.

DOUBLE PLAY
You're extremely likely to lose your appetite when you're upset if you both:
- don't tend to overeat at buffets
- would rather have a feeding tube than a colostomy bag

NO CORRELATION
People who lose their appetite when they're upset are almost exactly as likely as the average person to:
- have served as ring bearers or flower girls at a wedding
- take pleasure in putting things in order
- be sopranos or tenors

Macs vs. PCs

44% of people regularly back up their files.

But among those who prefer Macs over PCs, **61%** regularly back up their files.

78% of people like pesto sauce.

But among those who prefer Macs over PCs, **93%** like pesto sauce.

43% of people say they are the first result in a Google search of their name.

But among those who prefer Macs over PCs, **57%** say they are the first result in a Google search of their name.

DOUBLE PLAY

You're extremely likely to prefer Macs over PCs if you both:

- own an iPod or iPad
- use a laptop as your primary home computer

NO CORRELATION

People who prefer Macs over PCs are almost exactly as likely as the average person to:

- think the penny should not be done away with
- say they'd be willing to have all memories from the past year wiped out in exchange for $500,000
- be able to name more than one song by Flo Rida

Manual Transmission

68% of people know how to tie a necktie.

But among those who can't drive stick shift, only **52%** know how to tie a necktie.

38% of people tend to choose tails in a coin toss.

But among those who can't drive stick shift, **51%** tend to choose tails in a coin toss.

47% of people watch *The Big Bang Theory*.

But among those who can't drive stick shift, only **34%** watch *The Big Bang Theory*.

DOUBLE PLAY

You're extremely likely to be unable to drive stick shift if you both:

- are not an engineer
- are not better educated than your parents

NO CORRELATION

People who can't drive stick shift are almost exactly as likely as the average person to:

- prefer earbuds to headphones
- dislike pesto sauce
- say their favorite Girl Scout cookies are Thin Mints

Marijuana

33% of people have flown first class.

But among those who have smoked marijuana, **47%** have flown first class.

41% of people were into Nirvana.

But among those who have smoked marijuana, **55%** were into Nirvana.

44% of people are fans of actor Daniel Day-Lewis.

But among those who have smoked marijuana, **57%** are fans of actor Daniel Day-Lewis.

DOUBLE PLAY

You're extremely likely to have smoked marijuana if you both:

- have been given a hickey
- swear a lot

NO CORRELATION

People who have smoked marijuana are almost exactly as likely as the average person to:

- say they're more likely to buy the single than buy the album
- prefer silver-colored jewelry to gold-colored jewelry

people who trust professional critics' movie recommendations more than recommendations by their friends

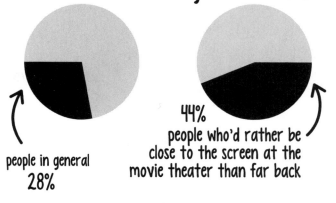

people in general
28%

44%
people who'd rather be close to the screen at the movie theater than far back

- rather find out they have a long-lost brother than a long-lost sister

Married People

48% of people have carved a turkey.

But among married people, **72%** have carved a turkey.

54% of people have donated blood.

But among married people, **70%** have donated blood.

49% of people snore.

But among married people, **65%** snore.

DOUBLE PLAY
You're extremely likely to be married if you both:
- have had hemorrhoids
- like to grill

NO CORRELATION
Non-married people are almost exactly as likely as the average person to:
- like Ernie more than Bert
- eat corn on the cob in a spiral
- get easily sunburned

Mayonnaise

56% of people think that if they could redo high school, knowing everything they know now, they would be just as uncool as they were the first time.

But among those who dislike mayonnaise, **67%** think that if they could redo high school, knowing everything they know now, they would be just as uncool as they were the first time.

43% of people find it difficult to explain to people what kind of work they do.

But among those who dislike mayonnaise, only **32%** find it difficult to explain to people what kind of work they do.

60% of people aren't good at remembering their dreams.

But among those who dislike mayonnaise, **70%** aren't good at remembering their dreams.

DOUBLE PLAY
You're extremely likely to dislike mayonnaise if you both:
- hate mushrooms
- have a high energy level

NO CORRELATION
People who dislike mayonnaise are almost exactly as likely as the average person to:
- prefer hard-shell tacos

- think their political opponents are, in general, more evil than stupid
- be social conservatives

Moonwalking

49% of people prefer creamy salad dressings to oily salad dressings.

But among those who know how to moonwalk, **62%** prefer creamy salad dressings.

60% of people would rather play a bass drum than a tuba in a marching band.

But among those who know how to moonwalk, **72%** would rather play a bass drum than a tuba in a marching band.

48% of people are skilled Capri Sun pouch puncturers.

But among those who know how to moonwalk, **60%** are skilled Capri Sun pouch puncturers.

DOUBLE PLAY

You're extremely likely to know how to moonwalk if you both:

- can spin a basketball on your finger
- like comedians who do celebrity impressions

NO CORRELATION

People who know how to moonwalk are almost exactly as likely as the average person to:

- plan their outfits the night before
- have visited someone in prison
- be faster than average readers

STATISTICS 101:
CORRELATION VS. ASSOCIATION

Technically, a correlation can exist only between two things that have a range of possible values, such as temperature or latitude. When there is a pattern to their relationship (for example, as one rises, so does the other, or as one rises, the other falls) we say that they are correlated.

The correlations in this book don't meet that formal definition because the surveys that were used asked people to choose from two opposing options, rather than a range of possible values. A statistics textbook would label them mere associations, rather than correlations.

Yet even though we can only informally use the term *correlation* to refer to these associations, this broader sense of the term is in widespread use, and although it may annoy some statistics sticklers, it's a meaning that's here to stay.

It's possible to have degrees of correlation and association, depending on how closely coupled the relationship is between the two variables.

For instance, you might find that there is a correlation, albeit a relatively weak one, between a person's annual income and his or her attractiveness, but a stronger correlation between a person's annual income and his or her education level.

Multilingualism

74% of people like mint chocolate chip ice cream.

But among those who are fluent in more than one language, only **56%** like mint chocolate chip ice cream.

39% of people type two spaces between sentences.

But among those who are fluent in more than one language, only **23%** type two spaces between sentences.

33% of people think Google hasn't done a good job of adhering to its unofficial "Don't be evil" motto.

But among those who are fluent in more than one language, **48%** think Google hasn't done a good job of adhering to its unofficial "Don't be evil" motto.

DOUBLE PLAY

You're extremely likely to be multilingual if you both:

- were not born in the United States
- don't find Australians sexy

NO CORRELATION

People who are multilingual are almost exactly as likely as the average person to:

- be side sleepers
- have had a root canal
- be skeptical about alternative medicine

Mushrooms

36% of people have never eaten a fig, except in Fig Newtons.

But among mushroom haters, **56%** have never eaten a fig, except in Fig Newtons.

47% of people prefer tea to coffee.

But among mushroom haters, **61%** prefer tea to coffee.

57% of people have never had athlete's foot.

But among mushroom haters, **69%** have never had athlete's foot.

DOUBLE PLAY

You're extremely likely to dislike mushrooms if you both:

- would not use a bidet, if available
- don't own a raincoat

NO CORRELATION

Mushroom haters are almost exactly as likely as the average person to:

- find it harder to forgive than ask for forgiveness
- ball up their socks to fold them
- have worn braces

Music Videos

46% of people think BASE jumpers are idiots.

But among those who think music videos are passé, **58%** think BASE jumpers are idiots.

49% of people say they feel like grown-ups.

But among those who think music videos are passé, **61%** say they feel like grown-ups.

36% of people think outdoor weddings aren't worth the risk of inclement weather.

But among those who think music videos are passé, **47%** think outdoor weddings aren't worth the risk of inclement weather.

DOUBLE PLAY

You're extremely likely to think music videos are passé if you both:

- have never purposely listened to a Justin Bieber song
- say your internal clock is pretty accurate

NO CORRELATION

People who think music videos are passé are almost exactly as likely as the average person to:

- be able to differentiate between a tangerine and a clementine
- be employed by the government
- have bad handwriting

Mustaches

60% of people stay in the movie theater until the credits end.

But among those who think men with mustaches are cool, **72%** stay in the movie theater until the credits end.

46% of people know who actress Felicia Day is.

But among those who think men with mustaches are cool, **57%** know who actress Felicia Day is.

60% of people like "Weird" Al Yankovic.

But among those who think men with mustaches are cool, **70%** like "Weird" Al Yankovic.

DOUBLE PLAY

You're extremely likely to think men with mustaches are cool if you both:

- are fluent in only one language
- think clowns are fun

NO CORRELATION

People who think men with mustaches are cool are almost exactly as likely as the average person to:

- have never been under general anesthesia
- be open to being a foster parent
- prefer to have a beautiful obituary over a beautiful gravestone

Nail Biting

48% of people are nervous pen clickers.

But among those who bite their nails, **59%** are nervous pen clickers.

52% of people regularly play video games.

But among those who bite their nails, **63%** regularly play video games.

68% of people would accept a friend's offer to let them hold a pet rat.

But among those who bite their nails, **78%** would accept a friend's offer to let them hold a pet rat.

DOUBLE PLAY
You're extremely likely to bite your nails if you both:
- describe yourself as absent-minded
- are typically believed to be older than you actually are

NO CORRELATION
People who bite their nails are almost exactly as likely as the average person to:
- say their right thumb is on the outside when they fold their hands
- be able to touch-type on a numerical keypad
- be employed by the government

Neckties

56% of people are good at making paper airplanes.

But among those who don't know how to tie a necktie, only **38%** are good at making paper airplanes.

32% of people prefer checkers to chess.

But among those who don't know how to tie a necktie, **49%** prefer checkers to chess.

50% of people snore.

But among those who don't know how to tie a necktie, only **35%** snore.

DOUBLE PLAY
You're extremely likely to not know how to tie a necktie if you both:
- bruise easily
- wear pajamas to bed

NO CORRELATION
People who don't know how to tie a necktie are almost exactly as likely as the average person to:
- spend more time listening to music than watching TV/movies
- say they would regift the Clapper
- like country music

Neighbors

31% of people are more likely to have an energy drink than a sports drink.

But among those who dislike their neighbors, **46%** are more likely to have an energy drink than a sports drink.

63% of people still make wishes before they blow out their birthday candles.

But among those who dislike their neighbors, only **49%** still make wishes before they blow out their birthday candles.

69% of people are good with kids.

But among those who dislike their neighbors, only **59%** are good with kids.

DOUBLE PLAY
You're extremely likely to dislike your neighbors if you both:
- are not very ticklish
- dislike licking envelopes

NO CORRELATION
People who dislike their neighbors are almost exactly as likely as the average person to:
- send out more e-cards than regular cards
- subscribe to Netflix
- prefer pi as the circle constant over tau

Nirvana

44% of people are delighted by modern art.

But among those who were into Nirvana, **58%** are delighted by modern art.

63% of people are well informed about world events.

But among those who were into Nirvana, **76%** are well informed about world events.

27% of people say their parents or grandparents are immigrants.

But among those who were into Nirvana, only **15%** say their parents or grandparents are immigrants.

people who prefer soft pillowcases

people in general
54%

69%
people who say their butt is the last thing they wash when they shower

DOUBLE PLAY

You're extremely likely to have been into Nirvana if you both:

- are a fan of the Foo Fighters
- are a fan of *Arrested Development*

NO CORRELATION

People who were into Nirvana are almost exactly as likely as the average person to:

- like skim milk
- say the person who most intimidates them is a woman
- have oily hair

Noncarbonated Beverages

39% of people make a lot of impulse purchases.

But among those who prefer noncarbonated beverages, only **28%** make a lot of impulse purchases.

47% of people bring their lunch more often than they buy their lunch.

But among those who prefer noncarbonated beverages, **56%** bring their lunch more often than they buy their lunch.

27% of people don't own flat-screen TVs.

But among those who prefer noncarbonated beverages, **37%** don't own flat-screen TVs.

You're extremely likely to prefer noncarbonated beverages if you both:

- prefer to sip cold beverages through a straw
- regularly eat yogurt

NO CORRELATION

People who prefer noncarbonated beverages are almost exactly as likely as the average person to:

- have made a prank call within the past three years
- dislike their nose
- regularly use a Laundromat

Non-College Graduates

45% of people aren't wine drinkers.

But among those who aren't college graduates, **64%** aren't wine drinkers.

62% of people prefer debit cards over credit cards.

But among those who aren't college graduates, **80%** prefer debit cards over credit cards.

47% of people have never gotten a speeding ticket.

But among those who aren't college graduates, **64%** have never gotten a speeding ticket.

DOUBLE PLAY

You're extremely likely to not be a college graduate if you both:

- don't have a nine-to-five type of job
- have been rickrolled

NO CORRELATION

People who are not college graduates are almost exactly as likely as the average person to:
- prefer their ice cream in a cone
- be able to tie a cherry stem into a knot with their tongue
- find freckled faces to be attractive

Nonfiction Lovers

46% of people like anime.

But among those who like nonfiction more than fiction, only **31%** like anime.

46% of people most enjoy living in urban areas.

But among those who like nonfiction more than fiction, **60%** most enjoy living in urban areas.

75% of people have a generally positive opinion about France.

But among those who like nonfiction more than fiction, **88%** have a generally positive opinion about France.

DOUBLE PLAY

You're extremely likely to prefer nonfiction over fiction if you both:

- have not read *The Hunger Games*
- say your kindergarten teacher would not have described you as self-reliant and resilient

NO CORRELATION

Nonfiction lovers are almost exactly as likely as the average person to:
- have eaten a fig, other than in Fig Newtons
- have broken a bone
- say they have very little shame

Noses

48% of people like techno music.

But among those who don't like their nose, only **36%** like techno music.

35% of people say that as children, they liked when things were neat and tidy.

But among those who don't like their nose, **47%** say that as children, they liked when things were neat and tidy.

37% of people use their middle finger to scroll when using a mouse with a scroll wheel.

But among those who don't like their nose, **49%** use their middle finger to scroll when using a mouse with a scroll wheel.

DOUBLE PLAY

You're extremely likely to dislike your nose if you both:

- have bad skin
- say the person you are in love with is not in love with you

NO CORRELATION

People who don't like their nose are almost exactly as likely as the average person to:

- have never pulled an all-nighter
- be unable to type without looking at the keyboard
- have been on blind dates

Older Siblings

43% of people spend more time watching TV/movies than listening to music.

But among those who have older siblings, **54%** spend more time watching TV/movies than listening to music.

55% of people are good at giving massages.

But among those who have older siblings, **64%** are good at giving massages.

42% of people have bouts of indigestion more often than bouts of indignation.

But among those who have older siblings, **50%** have bouts of indigestion more often than bouts of indignation.

DOUBLE PLAY

You're extremely likely to have older siblings if you both:

- think violent felons should not be made permanently ineligible to receive public food benefits
- think outdoor weddings are worth the risk of inclement weather

NO CORRELATION

People who have older siblings are almost exactly as likely as the average person to:

- have a Pinterest account
- prefer Leno to Letterman
- say they would get their son circumcised

On the Radio

35% of people have been involved in student government.

But among those who have been on the radio, **48%** have been involved in student government.

19% of people regularly tweet on Twitter.

But among those who have been on the radio, **31%** regularly tweet on Twitter.

13% of people have testified in court.

But among those who have been on the radio, **24%** have testified in court.

DOUBLE PLAY

You're extremely likely to have been on the radio if you both:

- have attended a wedding of a couple whom you didn't believe would last
- are a teacher

NO CORRELATION

People who have been on the radio are almost exactly as likely as the average person to:

- have been pantsed in public
- prefer soft pillowcases
- be able to recognize the quadratic formula if they saw it

Organ Donors

32% of people do the Electric Slide at wedding receptions.

But among those who aren't registered organ donors, only **19%** do the Electric Slide at wedding receptions.

24% of people say that white bread is their favorite kind of bread.

But among those who aren't registered organ donors, **35%** say that white bread is their favorite kind of bread.

62% of people want to be cremated.

But among those who aren't registered organ donors, only **51%** want to be cremated.

DOUBLE PLAY

You're extremely likely to not be a registered organ donor if you both:

- have never smoked marijuana
- never leave tips in tip jars

NO CORRELATION

People who aren't registered organ donors are almost exactly as likely as the average person to:

- say they'd rather find out they have a long-lost sister than a long-lost brother
- like super-cheesy pickup lines
- prefer Reese's Pieces to M&M's

Origami

68% of people would accept a friend's offer to let them hold a pet rat.

But among those who know how to make origami, **81%** would accept a friend's offer to let them hold a pet rat.

63% of people prefer tropical fruits to tropical climates.

But among those who know how to make origami, **75%** prefer tropical fruits to tropical climates.

40% of people say they don't have super-fast Internet connections.

But among those who know how to make origami, **49%** say they don't have super-fast Internet connections.

DOUBLE PLAY

You're extremely likely to know how to make origami if you both:

- have attempted to learn sign language
- are interested in geology

NO CORRELATION

People who know how to make origami are almost exactly as likely as the average person to:

- have extraordinary-sounding hiccups
- be surfers
- prefer a fine-mist shower stream rather than a high-pressure stream

Pants

73% of people know the difference between a stock and a bond.

But among those who prefer pleated pant fronts, **86%** know the difference between a stock and a bond.

72% of people like meat loaf dinners.

But among those who prefer pleated pant fronts, **85%** like meat loaf dinners.

41% of people would rather get a congratulatory stamp than a sticker.

But among those who prefer pleated pant fronts, **53%** would rather get a congratulatory stamp than a sticker.

DOUBLE PLAY

You're extremely likely to prefer pleated pant fronts if you both:

- don't get your hair cut at a salon
- believe in God

NO CORRELATION

People who prefer pleated pant fronts are almost exactly as likely as the average person to:

- say they are faster than average readers
- be good at remembering the words to songs
- be right-handed

people who would rather have the top bunk than the bottom bunk

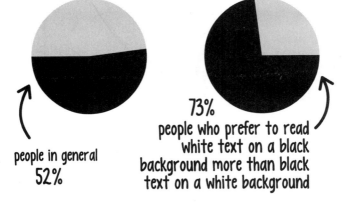

people in general
52%

73%
people who prefer to read white text on a black background more than black text on a white background

Parents

62% of people can drive stick shift.

But among those with children, **84%** can drive stick shift.

47% of people delete or move their emails after they've read them.

But among those with children, **65%** delete or move their emails after they've read them.

34% of people own an SLR camera.

But among those with children, **51%** own an SLR camera.

DOUBLE PLAY

You're extremely likely to be a parent if you both:

- have ridden a motorcycle
- think a loud baby is more tolerable than a loud dog

NO CORRELATION

People with children are almost exactly as likely as the average person to:

- have been on a swim team
- be willing to use a bidet
- prefer Coke over Pepsi

Passwords

43% of people are fans of Jim Carrey.

But among those who use the same password for pretty much everything, **55%** are fans of Jim Carrey.

38% of people would agree to staff a kissing booth to raise funds for their favorite charity.

But among those who use the same password for pretty much everything, **49%** would agree to staff a kissing booth to raise funds for their favorite charity.

61% of people say they remain objective and unbiased, even if it means potentially hurting someone's feelings.

But among those who use the same password for pretty much everything, only **51%** say they remain objective and unbiased, even if it means potentially hurting someone's feelings.

DOUBLE PLAY

You're extremely likely to use the same password for pretty much everything if you both:

- are more familiar with Play-Doh than Plato
- don't own an iPad or other tablet

NO CORRELATION

People who use the same password for pretty much everything are almost exactly as likely as the average person to:

- love *A Christmas Story*
- think the best way to respond to the Westboro Baptist Church's provocations is to ignore them, rather than counterprotest
- be blunt

People Who Dine Out Frequently

67% of people like detective fiction.

But among those who dine out frequently, **78%** like detective fiction.

37% of people are not very ticklish.

But among those who dine out frequently, **47%** are not very ticklish.

48% of people prefer a cologne/perfume that evokes autumn.

But among those who dine out frequently, **57%** prefer a cologne/perfume that evokes autumn.

DOUBLE PLAY

You're extremely likely to dine out frequently if you both:

- aren't bothered when different foods are touching on your plate
- have an American Express or Discover credit card

NO CORRELATION

People who dine out frequently are almost exactly as likely as the average person to:

- be lactose intolerant
- rather read the book than listen to the audiobook
- use shoe polish

STATISTICS 101:
THE MULTIPLE TESTING PROBLEM

Whenever you poll a group of people, there's always a chance that what seems like a significant result is actually just a matter of chance. For instance, if you flip a fair coin 100 times, it's highly unlikely that it would land on either heads or tails 61 or more times out of 100. In fact, there's only about a 5% chance of that happening. And so, we can say with 95% confidence (see page 19) that a fair coin will land on heads between 40 and 60 times in 100 flips.

Now suppose you're trying to determine whether a coin is fair and it lands on heads 61 times out of a hundred. You might conclude that it's biased toward heads.

But a 5% chance means that in 1 in 20 attempts, you should indeed expect to land on one side 61 or more times.

Thus the more times you conduct your coin-flipping experiment, the more likely it is that you will achieve what appears to be a significant result by pure chance.

Similarly, suppose you have data from hundreds of poll questions, and you compare the results of one poll to the results of all the other polls to see if any correlations exist. The more comparisons you make, the more likely it will be that an apparent correlation will occur by pure chance.

Pepsi vs. Coke

61% of people say they have a low energy level.

But among those who prefer Pepsi over Coke, **75%** say they have a low energy level.

47% of people don't feel comfortable opening their eyes underwater.

But among those who prefer Pepsi over Coke, **59%** don't feel comfortable opening their eyes underwater.

19% of people prefer powdered laundry detergent.

But among those who prefer Pepsi over Coke, only **8%** prefer powdered laundry detergent.

DOUBLE PLAY
You're extremely likely to prefer Pepsi over Coke if you both:

- are closer to lower class than upper class
- are good at keeping a straight face when you're pulling someone's leg

NO CORRELATION
People who prefer Pepsi over Coke are almost exactly as likely as the average person to:

- like techno music
- think Adam and Eve had belly buttons
- prefer crushed iced to cubed ice

Pessimists

28% of people would be willing to have all memories from the past year wiped out in exchange for $500,000.

But among pessimists, **44%** would be willing to have all memories from the past year wiped out in exchange for $500,000.

56% of people show their teeth when they smile for photos.

But among pessimists, only **42%** show their teeth when they smile for photos.

20% of people have never been bitten by an animal.

But among pessimists, **33%** have never been bitten by an animal.

DOUBLE PLAY

You're extremely likely to be a pessimist if you both:

- think children age you
- think the song title "Sometimes Love Just Ain't Enough" more closely represents your view than "All You Need Is Love"

NO CORRELATION

Pessimists are almost exactly as likely as the average person to:

- be interested in visiting the Salt and Pepper Shaker Museum in Gatlinburg, Tennessee
- want to take a martial arts class
- prefer scooped ice cream over soft serve

Pet Allergies

18% of people have food allergies.

But among those who are allergic to pet dander, **36%** have food allergies.

27% of people think it should be legal for two siblings to marry if they intend to not have children.

But among those who are allergic to pet dander, only **13%** think it should be legal for two siblings to marry if they intend to not have children.

43% of people occasionally use Craigslist.

But among those who are allergic to pet dander, **57%** occasionally use Craigslist.

DOUBLE PLAY

You're extremely likely to be allergic to pet dander if you both:

- regularly use ChapStick or other lip balm
- prefer to be photographed from your left side

NO CORRELATION

People who are allergic to pet dander are almost exactly as likely as the average person to:

- say they'd rather listen to the audiobook than read the book
- eat their chocolate bunnies butt-first
- know how to make origami

Philadelphia

59% of people like sea scallops.

But among those who would rather visit Philly for its cheesesteaks than for the Liberty Bell, **71%** like sea scallops.

47% of people prefer a cologne/perfume that evokes autumn rather than spring.

But among those who would rather visit Philly for its cheesesteaks than for the Liberty Bell, **58%** prefer a cologne/perfume that evokes autumn.

53% of people like punk rock.

But among those who would rather visit Philly for its cheesesteaks rather than for the Liberty Bell, **63%** like punk rock.

DOUBLE PLAY

You're extremely likely to prefer visiting Philly for its cheesesteaks rather than for the Liberty Bell if you both:

- think turducken sounds delicious
- don't have any parents or grandparents who are immigrants

NO CORRELATION

People who would rather visit Philly for its cheesesteaks than for the Liberty Bell are almost exactly as likely as the average person to:

- wear glasses or contacts
- prefer to dry their hands with an air dryer, rather than paper towels, in a public restroom
- prefer Mylar balloons

Piano Players

59% of people say they are better educated than their parents.

But among piano players, only **43%** are better educated than their parents.

43% of people consider themselves to be artistic.

But among piano players, **57%** consider themselves to be artistic.

60% of people prefer a combination lock for their locker rather than a lock with a key.

But among piano players, **72%** prefer a combination lock for their locker.

DOUBLE PLAY

You're extremely likely to play the piano if you both:
- have read the Chronicles of Narnia
- don't have a crown on any of your teeth

NO CORRELATION

Piano players are almost exactly as likely as the average person to:

- think it's OK to date your boss
- regularly wear cologne/perfume
- be left-handed

Pie vs. Cake

39% of people make a lot of impulse purchases.

But among those who prefer pie over cake, only **27%** make a lot of impulse purchases.

52% of people have read at least one gospel from the Bible all the way through.

But among those who prefer pie over cake, **62%** have read at least one gospel from the Bible all the way through.

69% of people like to grill.

But among those who prefer pie over cake, **78%** like to grill.

DOUBLE PLAY

You're extremely likely to prefer pie over cake if you both:

- had a great first kiss
- doodle infrequently

NO CORRELATION

People who prefer pie over cake are almost exactly as likely as the average person to:

- not have been a bed wetter
- hate *A Christmas Story*
- chew on their pens or pencils

Poetry

58% of people have eaten in a diner at 4 a.m.

But among those who prefer poetry that doesn't rhyme, **81%** have eaten in a diner at 4 a.m.

43% of people consider themselves to be artistic.

But among those who prefer poetry that doesn't rhyme, **56%** consider themselves to be artistic.

people who know who actress Felicia Day is

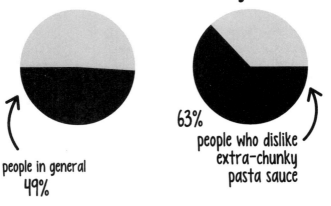

people in general
49%

63%
people who dislike
extra-chunky
pasta sauce

40% of people think Joe Biden is more likely than the Obama girls to one day end up on a reality show.

But among those who prefer poetry that doesn't rhyme, **53%** think Joe Biden is more likely than the Obama girls to one day end up on a reality show.

DOUBLE PLAY

You're extremely likely to prefer poetry that doesn't rhyme if you both:

- have read one of William Faulkner's works
- prefer jam to jelly

NO CORRELATION

People who prefer poetry that doesn't rhyme are almost exactly as likely as the average person to:

- not like to be alone when they're sick
- be unable to tell the difference between Coke and Pepsi
- prefer sausage over bacon on breakfast sandwiches

Poets

59% of people own thrift-store clothing.

But among those who write poetry, **73%** own thrift-store clothing.

19% of people would rather be blind than deaf.

But among those who write poetry, **32%** would rather be blind than deaf.

45% of people are homeowners.

But among those who write poetry, only **32%** are homeowners.

DOUBLE PLAY

You're extremely likely to be a poet if you both:

- meditate
- had a great first kiss

NO CORRELATION

People who write poetry are almost exactly as likely as the average person to:

- want a waterbed
- know the difference between amps, volts, and ohms
- like meat loaf dinners

Private School

39% of people say their mother was 30 or over when they were born.

But among those who attended private school, **54%** say their mother was 30 or over when they were born.

38% of people would agree to staff a kissing booth to raise funds for their favorite charity.

But among those who attended private school, **52%** would agree to staff a kissing booth to raise funds for their favorite charity.

36% of people think that a person should disclose that they are a widow/widower in their online dating profile.

But among those who attended private school, **50%** think that a person should disclose that they are a widow/widower in their online dating profile.

DOUBLE PLAY
You're extremely likely to have attended private school if you both:

- don't know how to use a sewing machine
- are the first result in a Google search of your name

NO CORRELATION
People who attended private school are almost exactly as likely as the average person to:

- have an uncommon first name
- prefer compact umbrellas over full-size
- be fast talkers

Privilege vs. Ambition

59% of people oppose capital punishment.

But among those who think privilege influences success more than ambition, **73%** oppose capital punishment.

74% of people can't pop a wheelie.

But among those who think privilege influences success more than ambition, **87%** can't pop a wheelie.

28% of people say they're pessimists.

But among those who think privilege influences success more than ambition, **38%** say they're pessimists.

DOUBLE PLAY

You're extremely likely to think privilege influences success more than ambition if you both:

- supported Barack Obama's reelection bid
- believe that churches and other places of worship should not be tax exempt

NO CORRELATION

People who think privilege influences success more than ambition are almost exactly as likely as the average person to:

- prefer an aisle seat on a plane
- pull the tabs off their soda cans
- prefer peanut M&M's over plain M&M's

Professional Wrestling

47% of people say acquaintances would describe them as soft spoken.

But among those who like professional wrestling, only **27%** say acquaintances would describe them as soft spoken.

50% of people would want to visit the Salt and Pepper Shaker Museum in Gatlinburg, Tennessee.

But among those who like professional wrestling, **66%** would want to visit the Salt and Pepper Shaker Museum in Gatlinburg, Tennessee.

66% of people prefer raw carrots over cooked.

But among those who like professional wrestling, **77%** prefer raw carrots over cooked.

DOUBLE PLAY

You're extremely likely to enjoy professional wrestling if you both:

- prefer that the open end of your pillowcase face the inside of the bed
- would eat a tasty-looking burger made from flesh cloned from your own body

NO CORRELATION

People who like professional wrestling are almost exactly as likely as the average person to:

- say they'd rather be a landscape photographer than a portrait photographer
- prefer *The Simpsons* to *South Park*
- prefer to shower in the morning

Public Speaking

38% of people describe themselves as passive-aggressive.

But among those who fear public speaking, **52%** describe themselves as passive-aggressive.

66% of people are better guessers than pantomimers when playing charades.

But among those who fear public speaking, **78%** are better guessers than pantomimers when playing charades.

52% of people have read at least one gospel from the Bible all the way through.

But among those who fear public speaking, **62%** have read at least one gospel from the Bible all the way through.

DOUBLE PLAY

You're extremely likely to fear public speaking if you both:

- say acquaintances would describe you as soft spoken
- are not a college graduate

NO CORRELATION

People who fear public speaking are almost exactly as likely as the average person to:

- prefer being called a nerd than a geek
- say that male pattern baldness runs in the family
- think the drinking age should be higher than the voting age

Restaurants and Bars

73% of people think the best way to respond to the Westboro Baptist Church's provocations is to ignore them, rather than counterprotest.

But among those who have worked in restaurants and bars, only **60%** think the best way to respond to the Westboro Baptist Church's provocations is to ignore them, rather than counterprotest.

81% of people have used a manual typewriter.

But among those who have worked in restaurants and bars, **93%** have used a manual typewriter.

39% of people think they would recognize the Dalai Lama, even if he walked past them in jeans and a Lakers jersey.

But among those who have worked in restaurants and bars, **50%** think they would recognize the Dalai Lama, even if he walked past them in jeans and a Lakers jersey.

DOUBLE PLAY
You're extremely likely to have worked in restaurants or bars if you both:
- have been cheated on
- are a fan of actor Daniel Day-Lewis

NO CORRELATION
People who have worked in restaurants and bars are almost exactly as likely as the average person to:

- rather act than direct
- get nosebleeds
- prefer bacon over sausage on breakfast sandwiches

Roller Coasters

50% of people say they're up for skydiving.

But among those who don't like roller coasters, only **23%** are up for skydiving.

62% of people prefer sandals to flip-flops.

But among those who don't like roller coasters, **75%** prefer sandals to flip-flops.

52% of people thought that it made sense when Pluto was demoted to dwarf planet status.

But among those who don't like roller coasters, **63%** thought that it made sense when Pluto was demoted to dwarf planet status.

DOUBLE PLAY

You're extremely likely to dislike roller coasters if you both:
- don't ski
- wash new clothes before they wear them

NO CORRELATION

People who don't like roller coasters are almost exactly as likely as the average person to:

- prefer to be the O in tic-tac-toe
- like mayonnaise
- be confused by modern art

Rolling Rs

50% of people don't like the smell of suntan lotion.

But among those who can't roll their *R*s, **61%** don't like the smell of suntan lotion.

54% of people know what their Myers-Briggs personality type is.

But among those who can't roll their *R*s, **65%** know what their Myers-Briggs personality type is.

26% of people don't tap their foot to the music.

But among those who can't roll their *R*s, **36%** don't tap their foot to the music.

DOUBLE PLAY

You're extremely likely to be unable to roll your *R*s if you both:

- are a better guesser than a pantomimer when playing charades
- like detective fiction

NO CORRELATION

People who can't roll their *R*s are almost exactly as likely as the average person to:

- prefer flavored water
- bite their nails
- think it'd be fun to have an identical twin

Sandwiches

49% of people had a great first kiss.

But among those who prefer their sandwiches cut into rectangles, **61%** had a great first kiss.

60% of people say they could never be in a relationship with a smoker.

But among those who prefer their sandwiches cut into rectangles, **70%** say they could never be in a relationship with a smoker.

24% of people like warm milk.

But among those who prefer their sandwiches cut into rectangles, only **14%** like warm milk.

DOUBLE PLAY

You're extremely likely to prefer your sandwiches cut into rectangles if you both:
- would rather fight 100 duck-size horses than one horse-size duck
- feel comfortable opening your eyes underwater

NO CORRELATION

People who prefer their sandwiches cut into rectangles are almost exactly as likely as the average person to:

- say they'd feel cheated out of birthdays if they were born on February 29
- study better with frequent breaks
- have dandruff

Scouts

33% of people have flown first class.

But among those who have been involved in scouting, **44%** have flown first class.

51% of people say their favorite kind of melon is not watermelon.

But among those who have been involved in scouting, **62%** say their favorite kind of melon is not watermelon.

61% of people have witnessed a meteor shower.

But among those who have been involved in scouting, **72%** have witnessed a meteor shower.

DOUBLE PLAY

You're extremely likely to have been involved in scouting if you both:

- have square danced
- have eaten venison

people who prefer Star Trek to Star Wars

people in general
43%

people who think ketchup
on eggs is yummy
58%

NO CORRELATION

People who have been involved in scouting are almost exactly as likely as the average person to:

- have step-parents
- be double-jointed
- prefer Superman over Batman

Sea-Monkeys

31% of people have been to a party that was broken up by the cops.

But among those who have tried to raise Sea-Monkeys, **43%** have been to a party that was broken up by the cops.

58% of people say their middle toe isn't longer than either adjacent toe.

But among those who have tried to raise Sea-Monkeys, **69%** say their middle toe isn't longer than either adjacent toe.

31% of people had an imaginary friend when they were a child.

But among those who have tried to raise Sea-Monkeys, **41%** had an imaginary friend when they were a child.

DOUBLE PLAY
You're extremely likely to have tried to raise Sea-Monkeys if you both:
- have had a urinary tract infection
- prefer crushed iced to cubed ice

NO CORRELATION
People who have tried to raise Sea-Monkeys are almost exactly as likely as the average person to:
- think the world would be better off if people were 10% dumber but 20% kinder
- belong to AAA
- be bothered by the sight of blood

Security vs. Liberty

26% of people prefer white beans over black beans.

But among those who value security over liberty, **49%** prefer white beans.

32% of people like Steven Tyler more than Liv Tyler.

But among those who value security over liberty, **47%** like Steven Tyler more than Liv Tyler.

49% of people have taken a martial arts class.

But among those who value security over liberty, only **35%** have taken a martial arts class.

DOUBLE PLAY

You're extremely likely to value security over liberty if you both:

- prefer cover bands with familiar songs to original music with unfamiliar songs
- have never voted for a minor party or write-in candidate

NO CORRELATION

People who value security over liberty are almost exactly as likely as the average person to:

- golf
- be more scared of snakes than spiders
- prefer Macs over PCs

Sense of Direction

78% of people say they aren't good at haggling.

But among those who have a bad sense of direction, **90%** aren't good at haggling.

45% of people say they'd rather have an indoor whirlpool tub than an outdoor hot tub.

But among those who have a bad sense of direction, **57%** would rather have an indoor whirlpool tub than an outdoor hot tub.

61% of people say they remain objective and unbiased, even if it means potentially hurting someone's feelings.

But among those who have a bad sense of direction, only **50%** say they remain objective and unbiased, even if it means potentially hurting someone's feelings.

DOUBLE PLAY

You're extremely likely to have a bad sense of direction if you both:

- are clumsier than the average person
- dislike sea scallops

NO CORRELATION

People who have a bad sense of direction are almost exactly as likely as the average person to:

- want to be cremated
- think climate change is caused by man
- think it's OK to text while peeing

Shower Time

25% of people tilt their head to the left when they go in for a kiss.

But among those who prefer to shower in the evening, **42%** tilt their head to the left when they go in for a kiss.

33% of people say that when squeezing past people who are seated, they point their crotch (rather than their butt) toward them.

But among those who prefer to shower in the evening, **46%** say that when squeezing past people who are seated, they point their crotch (rather than their butt) toward them.

43% of people are more interested in foosball than football.

But among those who prefer to shower in the evening, **56%** are more interested in foosball than football.

DOUBLE PLAY

You're extremely likely to prefer to shower in the evening if you both:

- have never been to a party that was broken up by the cops
- prefer *okay* over *OK*

NO CORRELATION

People who prefer to shower in the evening are almost exactly as likely as the average person to:

- be able to spin a basketball on their finger
- prefer poetry that rhymes
- not want to know the sex of a baby before it's born

STATISTICS 101: PRIMING

It's possible to influence the results of a poll simply by asking questions in a certain order or, in the case of multiple-choice questions, by listing the options in a certain order.

For instance, suppose you are conducting a political poll. You first ask, "What is our country's greatest problem?" Then you ask, "What is your opinion about the president?" People's answer to the first question is likely to influence their answer to the second question.

When pollsters intentionally order their questions in this way to influence responses, it's called priming, and for obvious reasons, it can affect the credibility of the poll.

To avoid any appearance of priming, most reputable polling companies randomize either the question order, the option order, or both.

Single-Sex Schools

81% of people think well-done graffiti is art.

But among those who like the idea of single-sex public schools, only **67%** think well-done graffiti is art.

28% of people have an American Express or Discover credit card.

But among those who like the idea of single-sex public schools, **41%** have an American Express or Discover credit card.

32% of people like country music.

But among those who like the idea of single-sex public schools, **44%** like country music.

DOUBLE PLAY
You're extremely likely to like the idea of single-sex public schools if you both:

- believe in objective morality
- have a long commute

NO CORRELATION
People who like single-sex public schools are almost exactly as likely as the average person to:

- like tomato soup
- have dated someone they met on a dating website
- have extraordinary-sounding hiccups

Skiers

48% of people say they make poorer choices when they're bored than when they're stressed.

But among skiers, **69%** say they make poorer choices when they're bored than when they're stressed.

52% of people aren't easily startled.

But among skiers, **73%** aren't easily startled.

55% of people claim to be good at parallel parking.

But among skiers, **73%** claim to be good at parallel parking.

DOUBLE PLAY
You're extremely likely to enjoy skiing if you both:
- have ridden on a zip line
- think ketchup on eggs is yummy

NO CORRELATION
Skiers are almost exactly as likely as the average person to:
- have done a fast or cleanse to remove toxins
- love bad puns
- chew gum frequently

Skydiving

53% of people think BASE jumpers are cool.

But among those who are up for skydiving, **71%** think BASE jumpers are cool.

80% of people have used a manual typewriter.

But among those who are up for skydiving, **91%** have used a manual typewriter.

67% of people say their internal clock is pretty accurate.

But among those who are up for skydiving, **77%** say their internal clock is pretty accurate.

DOUBLE PLAY

You're extremely likely to be up for skydiving if you both:

- like roller coasters
- think Halloween is more fun when you're of legal drinking age than when you're a kid

NO CORRELATION

People who are up for skydiving are almost exactly as likely as the average person to:

- have broken a bone
- smoke
- dislike sideburns

Slash vs. the Edge

31% of people are Justin Timberlake fans.

But among those who think the Edge is a better rock guitarist name than Slash, only **16%** are Justin Timberlake fans.

33% of people say their father was absent, estranged, or emotionally distant during their childhood.

But among those who think the Edge is a better rock guitarist name than Slash, only **21%** say their father was absent, estranged, or emotionally distant.

58% of people think old people are cute.

But among those who think the Edge is a better rock guitarist name than Slash, **67%** think old people are cute.

DOUBLE PLAY

You're extremely likely to think the Edge is a better rock guitarist name than Slash if you both:

- don't like cotton candy
- prefer *The Simpsons* to *South Park*

NO CORRELATION

People who think the Edge is a better rock guitarist name than Slash are almost exactly as likely as the average person to:

- have done a fast/cleanse to remove toxins
- like guacamole
- prefer non-mint gum

Sleep Position

55% of people consider themselves to be patriotic.

But among those who aren't side sleepers, **72%** consider themselves to be patriotic.

35% of people think that a person should disclose that they are a widow/widower in their online dating profile.

But among those who aren't side sleepers, **48%** think that a person should disclose that they are a widow/widower in their online dating profile.

56% of people are more interested in football than foosball.

But among those who aren't side sleepers, **68%** are more interested in football than foosball.

DOUBLE PLAY

You're extremely likely to not be a side sleeper if you both:

- assume that those without Facebook accounts are weird
- prefer a horizontal tire swing rather than a vertical tire swing

NO CORRELATION

People who aren't side sleepers are almost exactly as likely as the average person to:

- have deployed a fire extinguisher
- prefer fame over fortune
- have dimples

Smokers

67% of people have excellent credit.

But among those who smoke, only **36%** people have excellent credit.

42% of people have had a poison ivy rash.

But among those who smoke, **70%** have had a poison ivy rash.

41% of people prefer Reese's Pieces to M&M's.

But among those who smoke, **66%** prefer Reese's Pieces to M&M's.

DOUBLE PLAY

You're extremely likely to be a smoker if you both:

- are considered the black sheep of your family
- think Halloween is more fun when you're of legal drinking age than when you're a kid

NO CORRELATION

Smokers are almost exactly as likely as the average person to:

- suffer from canker sores
- prefer a cologne/perfume that evokes spring
- say their left hand does most of the work when they applaud

Sneaking In

56% of people never spill things on their keyboard.

But among those who have sneaked into a movie without paying, **75%** never spill things on their keyboard.

40% of people have ridden a pogo stick.

But among those who have sneaked into a movie without paying, **59%** have ridden a pogo stick.

80% of people would rather work four 10-hour shifts a week than five 8-hour shifts.

people who think Joe Biden is more likely than the Obama girls to one day end up on a reality show

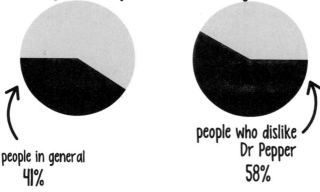

people in general
41%

people who dislike
Dr Pepper
58%

But among those who have sneaked into a movie without paying, **91%** would rather work four 10-hour shifts a week than five 8-hour shifts.

DOUBLE PLAY
You're extremely likely to have sneaked into a movie without paying if you both:
- have been on a bar crawl
- root for Wile E. Coyote over the Road Runner

NO CORRELATION
People who have sneaked into a movie without paying are almost exactly as likely as the average person to:
- think Holocaust denial should be illegal
- have been attacked by zombies
- tend to use dashes rather than slashes to delimit dates (for example, mm-dd-yy)

Soda vs. Pop

29% of people own tie-dyed clothing.

But among those who call cola "pop" rather than "soda," only **12%** own tie-dyed clothing.

63% of people say they are more familiar with Play-Doh than Plato.

But among those who call cola "pop" rather than "soda," **77%** say they are more familiar with Play-Doh than Plato.

46% of people say they have a tendency to push their luck.

But among those who call cola "pop" rather than "soda," only **34%** say they have a tendency to push their luck.

DOUBLE PLAY
You're extremely likely to call cola "pop" rather than "soda" if you both:
- like lamb
- prefer some other kind of bread besides white

NO CORRELATION
People who call cola "pop" rather than "soda" are almost exactly as likely as the average person to:
- not know how to make a PowerPoint presentation
- predominantly chew on the left side of their mouth
- dine out frequently

South Park vs. *The Simpsons*

37% of people would rather save up for an extravagant vacation than do a series of several more modest mini vacations.

But among those who prefer *South Park* to *The Simpsons*, **58%** would rather save up for an extravagant vacation than do a series of several more modest mini vacations.

66% of people say they're more likely to become vegetarian for health reasons than ethical reasons.

But among those who prefer *South Park* to *The Simpsons*, **80%** say they're more likely to become vegetarian for health reasons than ethical reasons.

40% of people have ridden a pogo stick.

But among those who prefer *South Park* to *The Simpsons*, **53%** have ridden a pogo stick.

DOUBLE PLAY
You're extremely likely to prefer *South Park* over *The Simpsons* if you both:
- think it'd be fun to have an identical twin
- would prefer to be corporeal, rather than incorporeal, if you could be invisible for a day

NO CORRELATION
People who prefer *South Park* to *The Simpsons* are almost exactly as likely as the average person to:
- prefer limp greens in their salad
- eat their chocolate bunnies ears-first
- prefer their corn off the cob

Speeding Tickets

67% of people aren't members of warehouse clubs.

But among those who have never gotten a speeding ticket, **79%** aren't members of warehouse clubs.

63% of people are fans of Penn & Teller.

But among those who have never gotten a speeding ticket, only **52%** are fans of Penn & Teller.

56% of people say their primary alarm clock is their phone.

But among those who have never gotten a speeding ticket, **67%** say their primary alarm clock is their phone.

DOUBLE PLAY
You're extremely likely to have never gotten a speeding ticket if you both:
- have never changed careers
- prefer that the open end of your pillowcase face the inside of the bed

NO CORRELATION
People who have never gotten a speeding ticket are almost exactly as likely as the average person to:
- dislike oregano
- like science fiction
- understand the word *conception* to mean fertilization

Spelling

55% of people consider themselves to be patriotic.

But among those who don't spell well, **71%** consider themselves to be patriotic.

38% of people ascend the stairs two at a time.

But among those who don't spell well, **53%** ascend the stairs two at a time.

29% of people think of soup as only a cold-weather food.

But among those who don't spell well, **45%** think of soup as only a cold-weather food.

DOUBLE PLAY
You're extremely likely to be a poor speller if you both:
- are fluent in only one language
- have never been prescribed a powerful painkiller

NO CORRELATION
People who don't spell well are almost exactly as likely as the average person to:
- like their current bank
- prefer leather couches over fabric couches
- say they can tolerate annoying people when it's in their own interest to do so

Spicy Food

26% of people prefer white beans to black beans.

But among those who don't like spicy food, **43%** prefer white beans.

51% of people have never carved a turkey.

But among those who don't like spicy food, **67%** have never carved a turkey.

58% of people have bad posture.

But among those who don't like spicy food, **73%** have bad posture.

DOUBLE PLAY

You're extremely likely to not like spicy food if you both:

- don't regularly exercise
- are not adept at using chopsticks

NO CORRELATION

People who don't like spicy food are almost exactly as likely as the average person to:

- think redheads are fascinating
- like warm milk
- prefer dodge ball over kickball

Spiders vs. Snakes

73% of people are college graduates.

But among those who are more scared of spiders than snakes, only **62%** are college graduates.

58% of people prefer reptiles to amphibians.

But among those who are more scared of spiders than snakes, **68%** prefer reptiles to amphibians.

55% of people use the phrase *a couple* only when referring to exactly two of something.

But among those who are more scared of spiders than snakes, **64%** use the phrase *a couple* only when referring to exactly two of something.

DOUBLE PLAY

You're extremely likely to be more scared of spiders than snakes if you both:

- prefer *Star Trek* to *Star Wars*
- have seen all three Lord of the Rings movies

NO CORRELATION

People who are more scared of spiders than snakes are almost exactly as likely as the average person to:

- have a hard time with *who* vs. *whom*
- say the person who most intimidates them is a woman
- say that if they had the opportunity to find out, they would want to know the date on which they will die

Step-Parents

53% of people have seen a therapist.

But among those who have step-parents, **71%** have seen a therapist.

22% of people say their favorite cereal brand is Kellogg's.

But among those who have step-parents, only **8%** say their favorite cereal brand is Kellogg's.

41% of people would rather get a congratulatory stamp than a sticker.

But among those who have step-parents, **53%** would rather get a congratulatory stamp than a sticker.

DOUBLE PLAY
You're extremely likely to have stepparents if you both:
- have worked in restaurants and bars
- tend to choose tails in a coin toss

NO CORRELATION
People who have stepparents are almost exactly as likely as the average person to:
- prefer the Summer Olympics over the Winter Olympics
- think flag burning should be legal
- own iPods/iPads

Stringed Instruments

27% of people can play the piano.

But among those who can play a stringed instrument, **47%** can play the piano.

38% of people like heavy metal.

But among those who can play a stringed instrument, **50%** like heavy metal.

69% of people prefer original music with unfamiliar songs to cover bands with familiar songs.

But among those who can play a stringed instrument, **80%** prefer original music with unfamiliar songs to cover bands with familiar songs.

DOUBLE PLAY

You're extremely likely to be able to play a stringed instrument if you both:

- would rather be blind than deaf
- enjoy camping

NO CORRELATION

People who can play a stringed instrument are almost exactly as likely as the average person to:

- prefer the radio/music alarm clock setting over the buzzer setting
- have sinus issues
- have taken a graduate-level admissions test

Subjective Morality

50% of people think a woman should take her husband's last name.

But among those who believe in subjective morality, only **30%** think a woman should take her husband's last name.

45% of people think there's been a net increase in morality over the past century.

But among those who believe in subjective morality, **62%** think there's been a net increase in morality over the past century.

82% of people support embryonic stem cell research.

But among those who believe in subjective morality, **96%** support embryonic stem cell research.

DOUBLE PLAY
You're extremely likely to believe in subjective morality if you both:
- consider yourself an atheist
- don't own an e-book reader

NO CORRELATION
People who believe in subjective morality are almost exactly as likely as the average person to:
- think Slash is a better rock guitarist name than the Edge
- like skim milk
- be good hula-hoopers

Surfers

32% of people say that when squeezing past people who are seated, they point their crotch (rather than their butt) toward them.

But among surfers, **53%** say that when squeezing past people who are seated, they point their crotch (rather than their butt) toward them.

61% of people own a bicycle.

But among surfers, **81%** own a bicycle.

57% of people would prefer to work at a place where they are considered a hotshot and are doted on rather than a place where they're in the middle of the pack.

But among surfers, **76%** would prefer to work at a place where they are considered a hotshot and are doted on rather than a place where they're in the middle of the pack.

people who show their teeth when they smile for photos

people in general
55%

people who own
tie-dyed clothing
74%

You're extremely likely to be a surfer if you both:

- have been on a bar crawl
- prefer Home Depot over Lowe's

NO CORRELATION

Surfers are almost exactly as likely as the average person to:

- have a negative opinion about homeschooling
- be light sleepers
- have up-to-date résumés

Sushi

48% of people dislike eggnog.

But among those who don't like sushi, **62%** dislike eggnog.

42% of people prefer milk chocolate to dark chocolate.

But among those who don't like sushi, **54%** prefer milk chocolate to dark chocolate.

69% of people like to be alone when they're sick.

But among those who don't like sushi, **79%** like to be alone when they're sick.

DOUBLE PLAY

You're extremely likely to not like sushi if you both:

- never leave tips in tip jars
- prefer your peas out of the pod

People who don't like sushi are almost exactly as likely as the average person to:

- think being Facebook friends with your boss is OK
- be bad at keeping secrets
- prefer electric toothbrushes

Sweet Snacks

52% of people don't regularly drink alcohol.

But among those who prefer sweet snacks over salty, **64%** don't regularly drink alcohol.

70% of people have never been kicked out of a public place.

But among those who prefer sweet snacks over salty, **80%** have never been kicked out of a public place.

38% of people regularly use an RSS reader.

But among those who prefer sweet snacks over salty, **47%** regularly use an RSS reader.

DOUBLE PLAY

You're extremely likely to prefer sweet snacks over salty if you both:

- have never smoked marijuana
- don't like pickles

People who prefer sweet snacks over salty are almost exactly as likely as the average person to:

- be more likely to flush than bury a dead fish
- be good at remembering lyrics
- like Dr Pepper

Tactfulness

40% of people prefer Reese's Pieces to M&M's.

But among those who describe themselves as blunt, **55%** prefer Reese's Pieces to M&M's.

49% of people deal with bullies by killing 'em with unkindness.

But among those who describe themselves as blunt, **64%** deal with bullies by killing 'em with unkindness.

34% of people think it's OK to text while peeing.

But among those who describe themselves as blunt, **45%** think it's OK to text while peeing.

DOUBLE PLAY

You're extremely likely to describe yourself as blunt if you both:

- just can't tolerate annoying people, even when it's in your own interest to do so
- prefer cherry cola to regular cola

People who are blunt are almost exactly as likely as the average person to:

- have attended private school
- say their left leg is on the bottom when they sit cross-legged
- most enjoy living in urban areas

Tattoos

67% of people like cauliflower.

But among those with tattoos, **90%** like cauliflower.

37% of people have worked in restaurants and bars.

But among those with tattoos, **60%** have worked in restaurants and bars.

40% of people went to a high school with a large marching band program.

But among those with tattoos, **59%** people went to a high school with a large marching band program.

DOUBLE PLAY
You're extremely likely to have tattoos if you both:
- 'occasionally use Craigslist
- have body piercings

NO CORRELATION

People with tattoos are almost exactly as likely as the average person to:

- have gotten straight *A*s
- support the "cry it out" approach to child sleep issues
- deal with bullies by killing 'em with kindness

Tea Drinkers

39% of people have never been given a hickey.

But among those who prefer tea to coffee, **50%** have never been given a hickey.

60% of people prefer mechanical pencils.

But among those who prefer tea to coffee, **71%** prefer mechanical pencils.

71% of people dislike licking envelopes.

But among those who prefer tea to coffee, **79%** dislike licking envelopes.

DOUBLE PLAY

You're extremely likely to prefer tea to coffee if you both:

- would support a smoking ban in all public places
- are more likely to have a sports drink than an energy drink

People who prefer tea to coffee are almost exactly as likely as the average person to:

- have a landline phone
- say that bellyaches are a bigger problem for them than headaches
- think they would be better off if they weren't so stubborn

Teachers

48% of people have carved a turkey.

But among teachers, **59%** have carved a turkey.

43% of people are homeowners.

But among teachers, **53%** are homeowners.

35% of people have read one of William Faulkner's works.

But among teachers, **44%** have read one of William Faulkner's works.

DOUBLE PLAY

You're extremely likely to be a teacher if you both:

- have used the word *obsequious* in conversation
- think that men who wear hats indoors are committing a faux pas

Teachers are almost exactly as likely as the average person to:

- like bananas
- think Will Smith is a better rapper than actor
- sympathize with celebrities who are hounded by paparazzi

Texting

62% of people prefer sandals to flip-flops.

But among those who don't frequently text, **74%** prefer sandals to flip-flops.

41% of people are cat lovers.

But among those who don't frequently text, **53%** are cat lovers.

49% of people are fast talkers.

But among those who don't frequently text, only **36%** are fast talkers.

DOUBLE PLAY
You're extremely likely to text infrequently if you both:

- don't have any posters on your bedroom wall
- don't prefer your vitamins in gummi bear form

NO CORRELATION

People who don't frequently text are almost exactly as likely as the average person to:

- regularly clip coupons
- have testified in court
- think Will Smith is a better rapper than actor

Theists

36% of people think couples should use traditional wedding vows.

But among those who believe in God, **55%** think couples should use traditional wedding vows.

57% of people admire Ronald Reagan more than Arnold Schwarzenegger.

people who like comedians who do celebrity impressions

people in general
55%

people who enjoy creamed chipped beef
71%

But among those who believe in God, **75%** admire Ronald Reagan more than Arnold Schwarzenegger.

63% of people have read the Chronicles of Narnia.

But among those who believe in God, **78%** have read the Chronicles of Narnia.

DOUBLE PLAY
You're extremely likely to believe in God if you both:
- believe the chicken came before the egg
- are involved in volunteer work

NO CORRELATION
People who believe in God are almost exactly as likely as the average person to:
- prefer a hot breakfast over a cold breakfast
- know how to moonwalk
- be good tree climbers

Thin People

55% of people say a nonphysical feature is the thing they like least about themselves.

But among thin people, **71%** say a nonphysical feature is the thing they like least about themselves.

32% of people say their father was absent, estranged, or emotionally distant.

But among thin people, only **18%** say their father was absent, estranged, or emotionally distant.

54% of people prefer the top bunk to the bottom bunk.

But among thin people, **64%** prefer the top bunk to the bottom bunk.

DOUBLE PLAY

You're extremely likely to describe yourself as thin if you both:

- thought that it made sense when Pluto was demoted to dwarf planet status
- don't want any (or any more) tattoos

NO CORRELATION

Thin people are almost exactly as likely as the average person to:

- have a positive opinion about fraternities/sororities
- think the robots of the future will be our masters, rather than our slaves
- say they would alert a stranger that his zipper was down

Tic-Tac-Toe

50% of people say that when playing a game with a child, they let the child win.

But among those who prefer to be the O in tic-tac-toe, **66%** let the child win.

24% of people would rather have a small meal and big dessert than a big meal and small dessert.

But among those who prefer to be the O in tic-tac-toe, **38%** would rather have a small meal and big dessert than a big meal and small dessert.

40% of people say they have oily hair.

But among those who prefer to be the O in tic-tac-toe, only **27%** people say they have oily hair.

DOUBLE PLAY
You're extremely likely to prefer to be the O in tic-tac-toe if you both:
- say your butt is the last thing you wash when you shower
- prefer charcoal grills over gas grills

NO CORRELATION
People who prefer to be the O in tic-tac-toe are almost exactly as likely as the average person to:
- think Delaware should cease to be a state and become a suburb of Philadelphia
- prefer poetry that rhymes
- have brown eyes

Time Travel

43% of people are homeowners.

But among those who would rather time-travel to the past than the future, **58%** are homeowners.

49% of people have read at least one gospel from the Bible all the way through.

But among those who would rather time-travel to the past than the future, **64%** have read at least one gospel from the Bible all the way through.

52% of people are more likely to do plumbing work on their own than electric work.

But among those who would rather time-travel to the past than the future, **60%** are more likely to do plumbing work on their own than electric work.

DOUBLE PLAY
You're extremely likely to say you'd rather time-travel to the past than the future if you both:

- admire Ronald Reagan more than Arnold Schwarzenegger
- don't regularly play video games

NO CORRELATION
People who would rather time-travel to the past than the future are almost exactly as likely as the average person to:

- get nosebleeds
- say their friends are mostly of the same sex
- prefer waffles to pancakes

Tipping a Server

39% of people have had a poison ivy rash.

But among those who refuse to tip a server when they receive poor service, only **18%** have had a poison ivy rash.

51% of people would not get their son circumcised.

But among those who refuse to tip a server when they receive poor service, **71%** would not get their son circumcised.

26% of people say the time 9:45 as "a quarter to ten" rather than "nine forty-five."

But among those who refuse to tip a server when they receive poor service, **42%** say the time 9:45 as "a quarter to ten" rather than "nine forty-five."

DOUBLE PLAY
You're extremely likely to refuse to tip a server when you receive poor service if you both:
- are more likely to sneeze into your hand than your sleeve
- have never taken a philosophy course

NO CORRELATION
People who refuse to tip a server when they receive poor service are almost exactly as likely as the average person to:
- say their preferred search engine is not Google
- have mistakenly used the public restroom of the opposite sex
- be able to juggle

Toilet Paper

67% of people believe it's their obligation to buy their designated driver's soft drinks.

But among those who prefer their toilet paper to hang against the wall, **89%** believe it's their obligation to buy their designated driver's soft drinks.

35% of people think Detroit would win in a fight against Chicago.

But among those who prefer their toilet paper to hang against the wall, **55%** think Detroit would win in a fight against Chicago.

41% of people say they have good posture.

But among those who prefer their toilet paper to hang against the wall, **59%** say they have good posture.

DOUBLE PLAY
You're extremely likely to prefer your toilet paper to hang against the wall if you both:
- prefer pistachios in the shell
- have attended a wedding of a couple whom you didn't believe would last

NO CORRELATION
People who prefer their toilet paper to hang against the wall are almost exactly as likely as the average person to:
- have attempted to learn sign language
- be allergic to pet dander
- have no interest in running for political office

Tongue Dexterity

46% of people have had pink eye.

But among those who can tie a cherry stem into a knot with their tongue, **67%** have had pink eye.

38% of people know how to make origami.

But among those who can tie a cherry stem into a knot with their tongue, **56%** know how to make origami.

62% of people have strong feelings about whether physician-assisted suicide should be legal.

But among those who can tie a cherry stem into a knot with their tongue, **78%** have strong feelings about whether physician-assisted suicide should be legal.

DOUBLE PLAY

You're extremely likely to be able to tie a cherry stem into a knot with your tongue if you both:

- prefer cherry cola to regular cola
- assume that those without Facebook accounts are weird

NO CORRELATION

People who can tie a cherry stem into a knot with their tongue are almost exactly as likely as the average person to:

- have thick hair
- wad up, rather than fold, their toilet paper
- like soggy cereal

Touch Typing

62% of people use a finger other than their middle finger to scroll when using a mouse with a scroll wheel.

But among those who can't type without looking at the keyboard, **79%** use a finger other than their middle finger to scroll when using a mouse with a scroll wheel.

58% of people have eaten in a diner at 4 a.m.

But among those who can't type without looking at the keyboard, **74%** have eaten in a diner at 4 a.m.

25% of people aren't fans of the serial comma.

But among those who can't type without looking at the keyboard, **39%** aren't fans of the serial comma.

DOUBLE PLAY

You're extremely likely to be unable to type without looking at the keyboard if you both:

- can't play the piano
- own flannel clothing

NO CORRELATION

People who can't type without looking at the keyboard are almost exactly as likely as the average person to:

- believe in objective morality
- use electric razors
- have dated someone of another race

TV in the Bedroom

73% of people think violent felons should not be made permanently ineligible to receive public food benefits.

But among those who have a TV in their bedroom, only **58%** think violent felons should not be made permanently ineligible to receive public food benefits.

45% of people have positive feelings about athletic scholarships.

But among those who have a TV in their bedroom, **59%** have positive feelings about athletic scholarships.

28% of people think North Korea's frequent threats of war are not empty threats.

But among those who have a TV in their bedroom, **40%** think North Korea's frequent threats of war are not empty threats.

DOUBLE PLAY

You're extremely likely to have a TV in your bedroom if you both:

- own an e-book reader
- use air freshener in your home

NO CORRELATION

People who have a TV in their bedroom are almost exactly as likely as the average person to:

- prefer their sandwiches cut into rectangles
- be Lady Gaga fans
- have sneaked into a movie without paying

Unemployment

50% of people have voted for a minor party or write-in candidate.

But among those who have collected unemployment, **67%** have voted for a minor party or write-in candidate.

28% of people have had hemorrhoids.

But among those who have collected unemployment, **45%** have had hemorrhoids.

32% of people prefer emoticons with noses.

But among those who have collected unemployment, **44%** prefer emoticons with noses.

DOUBLE PLAY
You're extremely likely to have collected unemployment if you both:
- have worked in the healthcare industry
- do your own taxes

NO CORRELATION
People who have collected unemployment are almost exactly as likely as the average person to:
- be more likely to bury than flush a dead fish
- prefer their gravy thick and gloopy rather than thin and soupy
- say that headaches are a bigger problem for them than bellyaches

Vegetarians

38% of people say they don't have very symmetrical faces.

But among vegetarians, **60%** say they don't have very symmetrical faces.

31% of people think the costs should be split during a date.

But among vegetarians, **52%** think the costs should be split during a date.

70% of people prefer jam to jelly.

But among vegetarians, **88%** prefer jam to jelly.

DOUBLE PLAY

You're extremely likely to be a vegetarian if you both:

- have a Tumblr account
- do yoga

NO CORRELATION

Vegetarians are almost exactly as likely as the average person to:

- like roller coasters
- prefer pancakes to waffles
- have collected unemployment

people who have tattoos

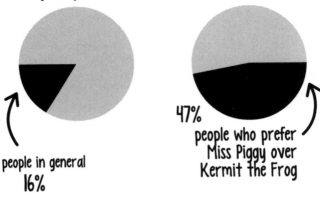

people in general
16%

47%
people who prefer Miss Piggy over Kermit the Frog

Video Games

46% of people know who actress Felicia Day is.

But among those who regularly play video games, **63%** know who actress Felicia Day is.

45% of people like anime.

But among those who regularly play video games, **59%** like anime.

62% of people understand HTML.

But among those who regularly play video games, **74%** understand HTML.

DOUBLE PLAY

You're extremely likely to regularly play video games if you both:

- like "Weird" Al Yankovic
- would rather time-travel to the future than the past

NO CORRELATION

People who regularly play video games are almost exactly as likely as the average person to:

- prefer French fries to onion rings
- prefer the term *firefly* over *lightning bug*
- be interested in genealogy

Volunteerism

30% of people are good singers.

But among those who are involved in volunteer work, **42%** are good singers.

58% of people think old people are cute.

But among those who are involved in volunteer work, **69%** think old people are cute.

54% of people know what their Myers-Briggs personality type is.

But among those who are involved in volunteer work, **64%** know what their Myers-Briggs personality type is.

DOUBLE PLAY

You're extremely likely to be involved in volunteer work if you both:

- know which directions "port" and "starboard" refer to
- can read music

NO CORRELATION

People who are involved in volunteer work are almost exactly as likely as the average person to:

- say they are more scared of bees than the average person
- think WD-40 is more useful than duct tape
- judge others based on how good-looking they are

Voting

63% of people say they're well informed about world events.

But among those who have voted for a minor party or write-in candidate, **76%** say they're well informed about world events.

29% of people own skinny jeans.

But among those who have voted for a minor party or write-in candidate, only **17%** own skinny jeans.

72% of people are not fastidious about locking the bathroom door.

But among those who have voted for a minor party or write-in candidate, **82%** are not fastidious about locking the bathroom door.

DOUBLE PLAY

You're extremely likely to have voted for a minor party or write-in candidate if you both:

- have been on the radio
- have boycotted a company

NO CORRELATION

People who have voted for a minor party or write-in candidate are almost exactly as likely as the average person to:

- like the smell of old books
- say they'd rather blow up at someone they're mad at than keep it in
- prefer Google Chrome as their browser

White Noise

26% of people say their parents or grandparents are immigrants.

But among those who prefer white noise when they sleep, only **14%** say their parents or grandparents are immigrants.

72% of people prefer their gravy thick and gloopy rather than thin and soupy.

But among those who prefer white noise when they sleep, **83%** prefer their gravy thick and gloopy rather than thin and soupy.

49% of people suffer from seasonal allergies.

But among those who prefer white noise when they sleep, **60%** suffer from seasonal allergies.

DOUBLE PLAY

You're extremely likely to prefer white noise when you sleep if you both:
- are better at punctuality than punctuation
- prefer cold weather over hot

NO CORRELATION

People who prefer white noise when they sleep are almost exactly as likely as the average person to:
- feel more confident in comfortable clothes than in dressy clothes
- know the difference between a stock and a bond
- have been on a cruise

Wisdom Teeth

50% of people have made a career change.

But among those who haven't had any wisdom teeth extracted, only **37%** have made a career change.

45% of people are not very physically affectionate.

But among those who haven't had any wisdom teeth extracted, **57%** are not very physically affectionate.

35% of people do not consider voting a moral duty.

But among those who haven't had any wisdom teeth extracted, **46%** do not consider voting a moral duty.

DOUBLE PLAY
You're extremely likely to have not had any of your wisdom teeth extracted if you both:

- have never taken a philosophy course
- like the smell of citronella

NO CORRELATION
People who haven't had any wisdom teeth extracted are almost exactly as likely as the average person to:

- be terrible bowlers
- believe women should be subject to the draft
- live alone

people who say the person who most intimidates them is a man

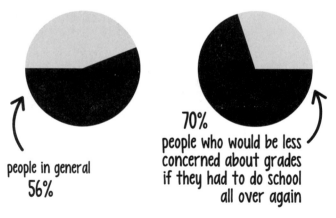

people in general
56%

70%
people who would be less concerned about grades if they had to do school all over again

Women

42% of people own cardigans.

But among women, **72%** own cardigans.

30% of people have nice handwriting.

But among women, **55%** have nice handwriting.

31% of people prefer checkers to chess.

But among women, **54%** prefer checkers to chess.

DOUBLE PLAY

You're extremely likely to be a woman if you both:

- prefer Hillary Clinton over Bill Clinton
- prefer to sip cold beverages through a straw

NO CORRELATION

Women are almost exactly as likely as the average person to:

- think fireworks displays are better when they're set to music
- have used the word *obsequious* in conversation
- find Australians sexy

Work Less vs. Earn More

59% of people are dog lovers.

But among those who would rather work less than earn more, only **45%** are dog lovers.

54% of people know what their Myers-Briggs personality type is.

But among those who would rather work less than earn more, **66%** know what their Myers-Briggs personality type is.

37% of people don't make wishes before they blow out their birthday candles.

But among those who would rather work less than earn more, **48%** don't make wishes before they blow out their birthday candles.

DOUBLE PLAY

You're extremely likely to say you'd rather work less than earn more if you both:

- tend to choose tails in a coin toss
- dislike Big Business more than Big Government

NO CORRELATION

People who would rather work less than earn more are almost exactly as likely as the average person to:

- regularly check their horoscope
- plan their outfits the night before
- have gone glow bowling

World War III

54% of people think there's been a net decrease in morality over the past century.

But among those who think World War III will occur in the next 25 years, **68%** think there's been a net decrease in morality over the past century.

38% of people say they are passive-aggressive.

But among those who think World War III will occur in the next 25 years, **52%** say they are passive-aggressive.

56% of people prefer *Star Wars* to *Star Trek*.

But among those who think World War III will occur in the next 25 years, **68%** prefer *Star Wars* to *Star Trek*.

DOUBLE PLAY

You're extremely likely to think World War III will occur in the next 25 years if you both:

- dislike the scent of menthol
- would regift the Clapper

NO CORRELATION

People who think World War III will occur in the next 25 years are almost exactly as likely as the average person to:

- have summer birthdays
- prefer their sandwiches cut into triangles
- have taken a martial arts class

Yoga

32% of people prefer sweet potatoes to potatoes.

But among those who do yoga, **53%** prefer sweet potatoes to potatoes.

27% of people regularly use mouthwash.

But among those who do yoga, **44%** regularly use mouthwash.

22% of people have a Pinterest account.

But among those who do yoga, **38%** have a Pinterest account.

DOUBLE PLAY
You're extremely likely to do yoga if you both:
- are a Lady Gaga fan
- like having your feet massaged

NO CORRELATION
People who do yoga are almost exactly as likely as the average person to:
- say they'd rather be called a nerd than a geek
- like their current bank
- be Caucasian

Zombie Victims

23% of people have Tumblr accounts.

But among those who say they have been attacked by zombies, **55%** have Tumblr accounts.

48% of people say their interest in sports is lower than their ability to play sports.

But among those who say they have been attacked by zombies, **70%** say their interest in sports is lower than their ability to play sports.

37% of people prefer their vitamins in gummi bear form.

But among those who say they have been attacked by zombies, **55%** prefer their vitamins in gummi bear form.

DOUBLE PLAY
You're extremely likely to have been attacked by zombies if you both:
- prefer to dry your hands with air dryers, rather than paper towels, in public restrooms
- prefer Batman over Superman

NO CORRELATION
People who have been attacked by zombies are almost exactly as likely as the average person to:
- prefer foreign cars
- describe themselves as a little bit racist
- have served as ring bearers or flower girls at a wedding

ACKNOWLEDGMENTS

Foremost thanks go to Correlated.org's contributors, who supplied the data used in this book, and especially to the top contributors—numbering several hundred—who have answered more than 90% of the poll questions that have been published since Correlated's inception.

Thanks also to Marian Lizzi and Maria Gagliano at Perigee, Laurie Abkemeier at DeFiore & Co., and to everyone else involved in the production of the book.

Finally, I'd like to thank my wife, Tanya, and the rest of my family. Their support, encouragement, and feedback have been instrumental in bringing this book to life. They have celebrated my strengths and put up with my many weaknesses, and for that I am deeply grateful.

ABOUT THE AUTHOR

Shaun Gallagher is the founder of Correlated.org and is also the author of *Experimenting with Babies: 50 Amazing Science Projects You Can Perform on Your Kid* (experimentingwithbabies.com).